The ESSENTIALS of

Calculus I

Staff of Research and Education Association
Dr. M. Fogiel, Director

This book covers the usual course outline of Calculus I. For more advanced topics, see *"THE ESSENTIALS OF CALCULUS II"* and *"THE ESSENTIALS OF CALCULUS III."*

Research & Education Association
61 Ethel Road West
Piscataway, New Jersey 08854

THE ESSENTIALS®
OF CALCULUS I

Year 2001 Printing

Printed in the United States of America

Library of Congress Control Number 99-76836

International Standard Book Number 0-87891-577-X

IV-2

WHAT "THE ESSENTIALS" WILL DO FOR YOU

This book is a review and study guide. It is comprehensive and it is concise.

It helps in preparing for exams and in doing homework, and remains a handy reference source at all times.

It condenses the vast amount of detail characteristic of the subject matter and summarizes the **essentials** of the field.

It will thus save hours of study and preparation time.

The book provides quick access to the important facts, principles, theorems, concepts, and equations of the field.

Materials needed for exams can be reviewed in summary form— eliminating the need to read and re-read many pages of textbook and class notes. The summaries will even tend to bring detail to mind that had been previously read or noted.

This "ESSENTIALS" book has been prepared by experts in the field, and has been carefully reviewed to ensure accuracy and maximum usefulness.

Dr. Max Fogiel
Program Director

CONTENTS

This book covers the usual course outline of Calculus I. For more advanced topics, see *"THE ESSENTIALS OF CALCULUS II"* and *"THE ESSENTIALS OF CALCULUS III"*.

CHAPTER 1

FUNDAMENTALS

1.1 NUMBER SYSTEMS

The real number system can be broken down into several parts and each of these parts have certain operations which can be performed on them. First, let us define the components of the real number system.

The natural numbers, denoted N, are 1,2,3,4,.... The integers, denoted Z, are ...-3,-2,-1,0,1,2,3,.... The rational numbers, denoted Q, are all numbers of the form p/q where p and q are integers and $q \neq 0$. A real number x is a non-terminating decimal (with a sign + or -).

Six basic algebraic properties of rational numbers:

a) The closure property: If x and y are rational numbers, then $x+y$ and $x \cdot y$ are also rational numbers.

b) Additive and multiplicative identity elements: If x is a rational number, then $x+0=x$ and $x \cdot 1=x$.

c) Associative property: If x, y and z are rational numbers, then $x+(y+z)=(x+y)+z$, $x(y \cdot z)=(x \cdot y)z$.

d) Additive and multiplicative inverses: For each rational number, x, such that $x+(-x)=0$; if $x \neq 0$, there exists a rational number x^{-1} such that $x \cdot x^{-1}=1$.

e) Commutative property: If x and y are rational numbers, then $x+y=y+x$, $x \cdot y=y \cdot x$.

f) Distributive property: If x, y, and z are rational numbers, then

$$x \cdot (y+z) = (x \cdot y) + (x \cdot z)$$

If q and p are rational numbers and $p-q$ is negative, then q is greater than p, $(q>p)$ or p is less than q, $(p<q)$.

1.1.1 PROPERTIES OF RATIONAL NUMBERS

a) Trichotomy property: If p and q are rational numbers, then one and only one of the relations $q=p$, $q>p$ or $q<p$ is true.

b) Transitive property: If p, q, and r are rational numbers, and if $p<q$ and $q<r$, then $p<r$.

c) If p, q, and r are rational numbers and $p<q$, then $p+r<q+r$.

d) If p, q, and r are rational numbers and if $p<q$ and $0<r$, then $pr<qr$.

1.2 INEQUALITIES

To solve a linear inequality

$$ax+b>0 \text{ or } x>-b/a, \text{ where } a>0$$

draw a number line, dashed for $x<-b/a$ and solid for $x>-b/a$.

$\frac{-b}{a}$ x

To solve $(ax+b)(cx+d)>0$ graphically, where $a>0$ and $c>0$:

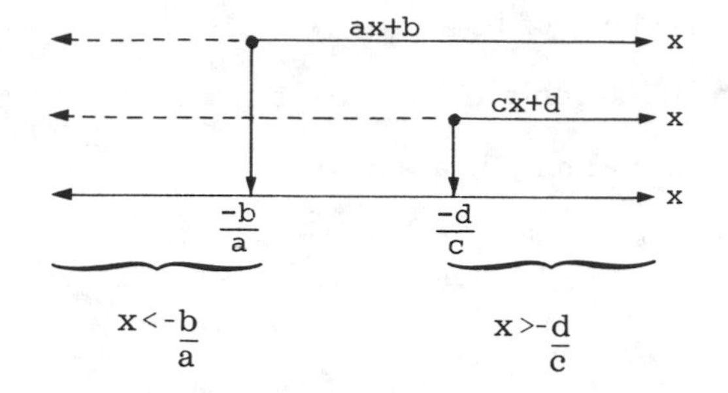

The solution lies in the interval where both lines are dashed and both lines are solid.

Hence, $x>-d/c$, $x<\frac{-b}{a}$ is the solution to the above inequality.

1.3 ABSOLUTE VALUE

Definition: The absolute value of a real number x is defined as

$$|x| = \begin{cases} x \text{ if } x \geq 0 \\ -x \text{ if } x<0 \end{cases}$$

For real numbers a and b:

a) $|a| = |-a|$

b) $|ab| = |a| \cdot |b|$

c) $-|a| \leq a \leq |a|$

d) $ab \leq |a||b|$

e) $|a+b|^2 = (a+b)^2$

f) $|a+b| \leq |a| + |b|$ (Triangle Inequality)

g) $|a-b| \geq ||a| - |b||$

For positive values of b

a) $|a| < b$ if and only if $-b < a < b$

b) $|a| > b$ if and only if $a > b$ or $a < -b$

c) $|a| = b$ if and only if $a = b$ or $a = -b$

1.4 SET NOTATION

A set is a collection of objects called elements. Let A and B be sets.

$x \in A$: x is an element of A
$x \notin B$: x is not an element of B

A is a subset of B, $(A \subset B)$, means that A is contained in another set B and each element of A is also an element of B.

A is equal to B $(A=B)$, if and only if $A \subset B$ and $B \subset A$.

$A \cup B$: The union of A and B; the set consists of all elements of A and B.

$A \cap B$: the intersection of A and B; the set consists of elements, common to both A and B.

$A \cap B = \phi$: It is the set that has no elements common to both A and B; thus it is an empty set. Empty sets are said to be disjoint.

These notations may be used to describe intervals of numbers such as:

The open interval $(a,b) = \{x : a < x < b\}$
The closed interval $[a,b] = \{x : a \le x \le b\}$
The half-open intervals $[a,b) = \{x : a \le x < b\}$
and $(a,b] = \{x : a < x \le b\}$

1.5 SUMMATION NOTATION

If we are given a set or collection of numbers $\{a_1, a_2, a_3, \ldots, a_n\}$, the sum of these numbers can be represented by the symbol $\sum_{i=1}^{n} a_i$, that is,

$$\sum_{i=1}^{n} a_i = a_1 + a_2 + a_3 + \ldots + a_n.$$

In general, $\sum_{i=1}^{n} c = nc$ for every real number c, and

$$\sum_{i=1}^{n} c\, a_i = c\left(\sum_{i=1}^{n} a_i\right).$$

For any positive integer n and the sets of numbers $\{a_1, a_2, a_3, \ldots, a_n\}$ and $\{b_1, b_2, b_3, \ldots, b_n\}$,

$$\sum_{i=1}^{n} (a_i + b_i) = \sum_{i=1}^{n} a_i + \sum_{i=1}^{n} b_i$$

$$\sum_{i=1}^{n} (a_i - b_i) = \sum_{i=1}^{n} a_i - \sum_{i=1}^{n} b_i$$

1.6 MATHEMATICAL INDUCTION

If with each positive integer n, there is associated a statement or proposition P_n to be proven, then all the statements P_n are true provided the following conditions hold:

1) P_1 is true
2) Whenever k is a positive integer such that P_k is true, then P_{k+1} is also true.

To apply this principle:

Step 1. Prove that P_1 is true.
Step 2. Assume that P_k is true and prove from this that P_{k+1} is also true.

CHAPTER 2

FUNCTIONS

2.1 FUNCTIONS

Definition: function is a correspondence between two sets, the domain and the range, such that for each value in the domain there corresponds exactly one value in the range.

A function has three distinct features:

a) the set x which is the domain,

b) the set y which is the co-domain or range,

c) a functional rule, f, that assigns only one element $y \in Y$ to each $x \in X$. We write $y = f(x)$ to denote the functional value y at x.

Consider Figure 2.1. The "machine" f transforms the domain X, element by element, into the co-domain Y.

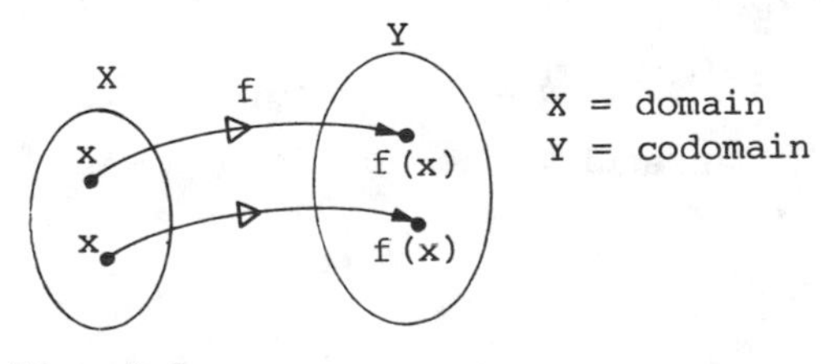

Fig. 2.1

2.2 COMBINATION OF FUNCTIONS

Let f and g represent functions, then

a) the sum $(f+g)(x) = f(x) + g(x)$,
b) the difference $(f-g)(x) = f(x) - g(x)$,
c) the product $(fg)(x) = f(x)g(x)$,

d) the quotient $(\frac{f}{g})(x) = \frac{f(x)}{g(x)}$, $g(x) \neq 0$,

e) the composite function $(g \circ f)(x) = g(f(x))$ where $f(x)$ must be in the domain of g.

A polynomial function of degree n is denoted as

$$f(x) = a_n x^n + a_{n-1} x^{n-1} + a_{n-2} x^{n-2}$$

$$+ \ldots + a_1 x + a_0$$

where a_n is the leading coefficient and not equal to zero

and $a_k x^k$ is the kth term of the polynomial.

2.3 PROPERTIES OF FUNCTIONS

A) A function F is one to one if for every range value there corresponds exactly one domain value of x.

B) A function is even if $f(-x) = f(x)$ or

$$f(x) + f(-x) = 2f(x).$$

C) A function is said to be odd if $f(-x) = -f(x)$ or $f(x) + f(-x) = 0$.

D) Periodicity

A function f with domain X is periodic if there exists a

positive real number p such that $f(x+p) = f(x)$ for all $x \in X$.

The smallest number p with this property is called the period of f.

Over any interval of length p, the behavior of a periodic function can be completely described.

E) Inverse of a function

Assuming that f is a one-to-one function with domain X and range Y, then a function g having domain Y and range X is called the inverse function of f if:

$f(g(y)) = y$ for every $y \in Y$ and

$g(f(x)) = x$ for every $x \in X$.

The inverse of the function f is denoted f^{-1}.

To find the inverse function f^{-1}, you must solve the equation $y = f(x)$ for x in terms of y.

Be careful: This solution must be a function.

F) The identity function $f(x) = x$ maps every x to itself.

G) The constant function $f(x) = c$ for all $x \in R$.

The "zeros" of an arbitrary function f(x) are particular values of x for which $f(x) = 0$.

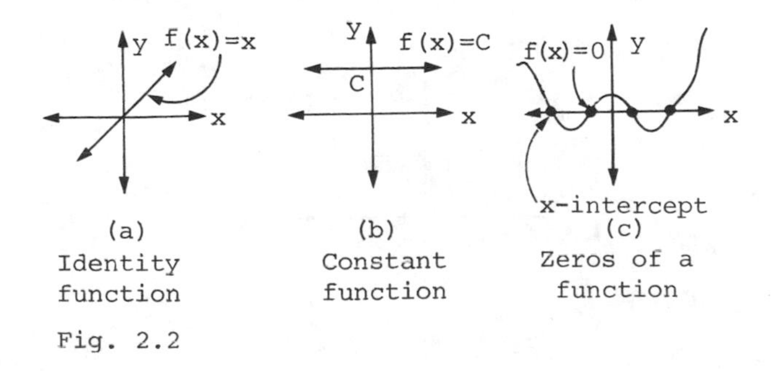

(a) Identity function

(b) Constant function

(c) Zeros of a function

Fig. 2.2

2.4 GRAPHING A FUNCTION

2.4.1 THE CARTESIAN COORDINATE SYSTEM

Consider two lines x and y drawn on a plane region called R.

Let the intersection of x and y be the origin and let us impose a coordinate system on each of the lines.

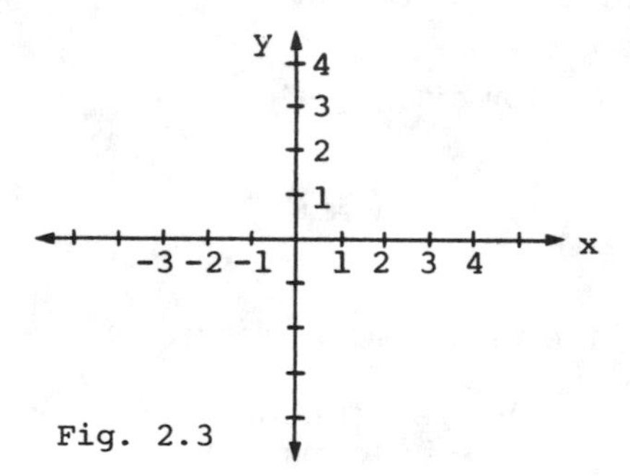

Fig. 2.3

If (x,y) is a point or ordered pair on the coordinate plane R then x is the first coordinate and y is the second coordinate.

To locate an ordered pair on the coordinate plane simply measure the distance of x units along the x-axis, then measure vertically (parallel to the y-axis) y units.

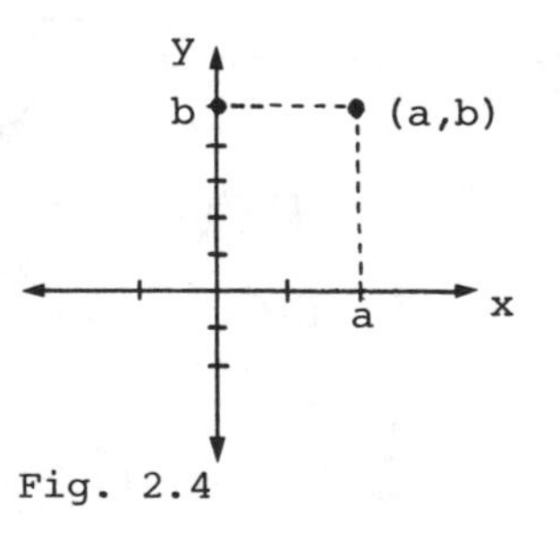

Fig. 2.4

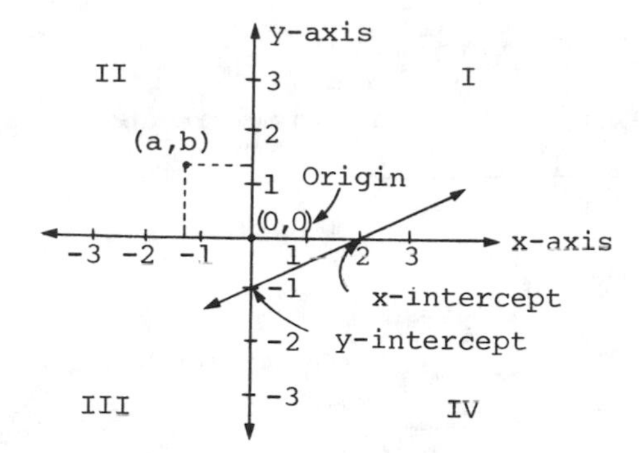

I, II, III, IV are called quadrants in the COORDINATE PLANE.

(a,b) is an ordered pair with x-coordinate a and y-coordiante b.

Fig. 2.5 CARTESIAN COORDINATE SYSTEM

2.4.2 DRAWING THE GRAPH

There are several ways to plot the graph of a function. The process of computing and plotting points on the graph is always an aid in this endeavor. The more points we locate on the graph, the more accurate our drawing will be.

It is also helpful if we consider the symmetry of the function. That is,

a) A graph is symmetric with respect to the x-axis if whenever a point (x,y) is on the graph, then (x,-y) is also on the graph.

b) Symmetry with respect to the y-axis occurs when both points (-x,y) and (x,y) appear on the graph for every x and y.

c) When the simultaneous substitution of -x for x and -y for y does not change the solution of the equation, the graph is said to be symmetric about the origin.

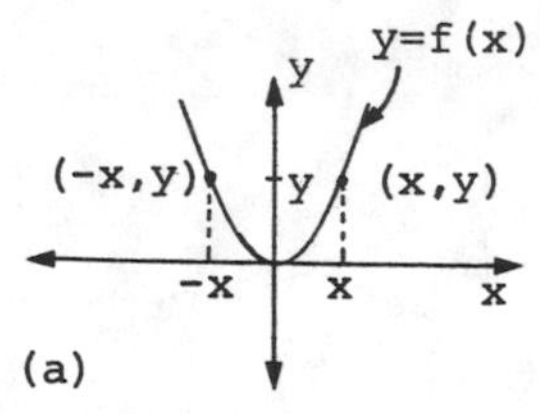

Symmetric about the y-axis

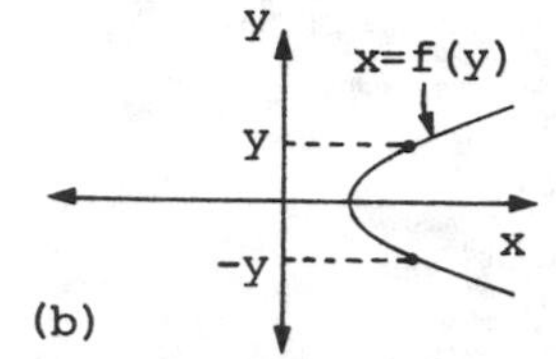

Symmetric about the x-axis

Note: This is not a function of x.

(c)

Symmetric about the origin

Fig. 2.6

Another aid in drawing a graph is locating any vertical asymptotes.

A vertical asymptote is a vertical line $x = a$, such that the functional value $|f(x)|$ grows indefinitely large as x approaches the fixed value a.

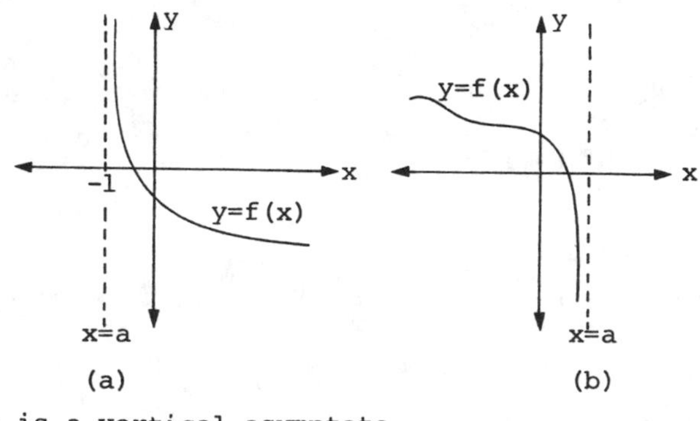

x=a is a vertical asymptote for this function

Fig. 2.7

The following steps encapsulate the procedure for drawing a graph:

a) Determine the domain and range of the function.
b) Find the intercepts of the graph and plot them.
c) Determine the symmetries of the graph.
d) Locate the vertical asymptotes and plot a few points on the graph near each asymptote.
e) Plot additional points as needed.

2.5 LINES AND SLOPES

Each straight line in a coordinate plane has an equation of the form $Ax + By + C = 0$, where A and B are not zero.

If we consider only a portion or a segment of the line we can find both, the length of the segment and its midpoint.

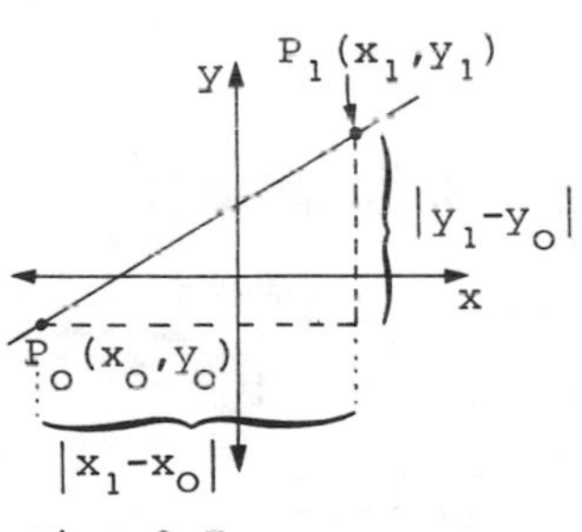

Fig. 2.8

The distance between two points P_0 and P_1 in a coordinate plane is $d(P_0,P_1) = \sqrt{(x_1-x_0)^2+(y_1-y_0)^2}$.

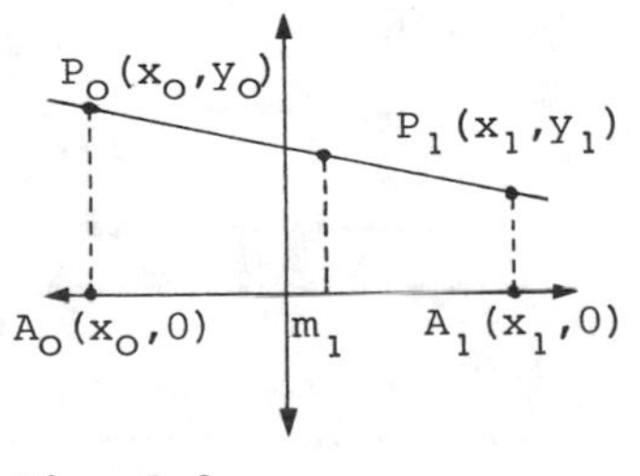

Fig. 2.9

The midpoint of a line segment from P_0 to P_1 is the point

$$\left(\frac{x_1+x_0}{2}, \frac{y_1+y_0}{2}\right)$$

However we are more often concerned with finding the slope of the line.

If given two points (x_1,y_1) and (x_0,y_0) the ratio

$$\frac{y_1-y_0}{x_1-x_0}$$

is the slope of the line.

Any two segments of the same line must have the same slope. Therefore looking at Fig. 2.10 we see

$$\frac{y_3-y_2}{x_3-x_2} = \frac{y_1-y_0}{x_1-x_0}.$$

It is easy to show that if two line segments have the same slopes and a common endpoint, then they must be the same line.

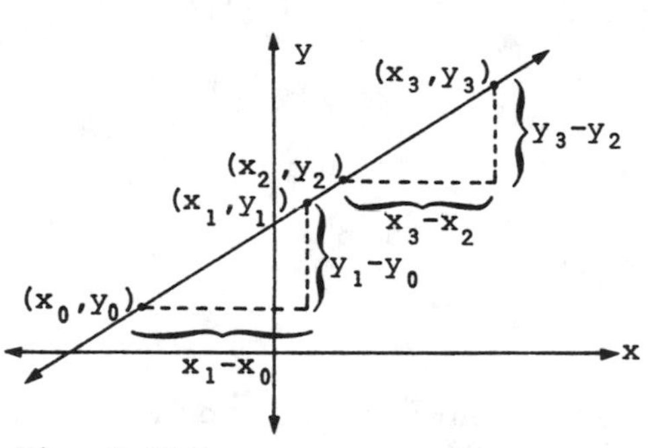

Fig. 2.10

The equation for a line can be conveniently written as

$$\boxed{y = mx + b}$$

where $$\boxed{m = \text{slope} = \frac{y_1-y_0}{x_1-x_0}}$$

and b = y-intercept; where the line intersects the y-axis.

The value of m will help us determine the position of the line on a graph.

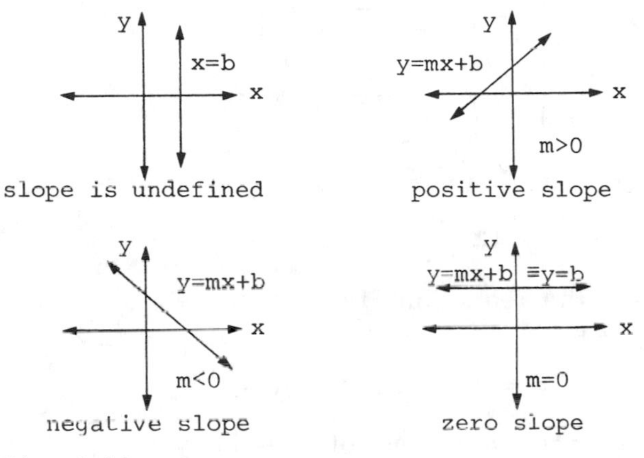

Fig. 2.11

The slope of a line can be used to determine whether or not several points are collinear. Given n points (a_1,b_1) $(a_2,b_2),\ldots,(a_n,b_n)$ they are collinear if and only if

$$\frac{b_i-b_{i-1}}{a_i-a_{i-1}}=\frac{b_2-b_1}{a_2-a_1} \quad \text{for } i = 3,4,\ldots n.$$

Two lines are parallel if and only if their slopes are equal.

Two lines having slopes m_1 and m_2 are perpendicular if and only if $m_1m_2 = -1$.

2.6 PARAMETRIC EQUATIONS

If we have an equation y = f(x), and the explicit functional form contains an arbitrary constant called a parameter, then it is called a parametric equation. A function with a parameter represents not one but a family of curves.

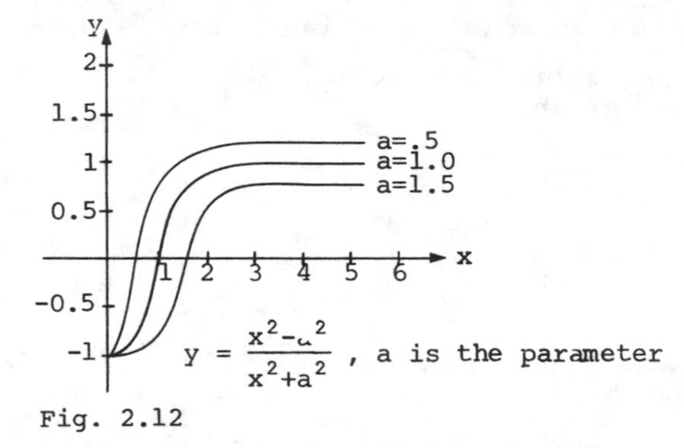

Fig. 2.12

Often the equation for a curve is given as two functions of a parameter t, such as

$$X = x(t) \text{ and } Y = y(t).$$

Corresponding values of x and y are calculated by solving for t and substituting.

CHAPTER 3

TRANSCENDENTAL FUNCTIONS

3.1 TRIGONOMETRIC FUNCTIONS

The trigonometric functions are defined in terms of a point P which moves in a circular track of unit radius.

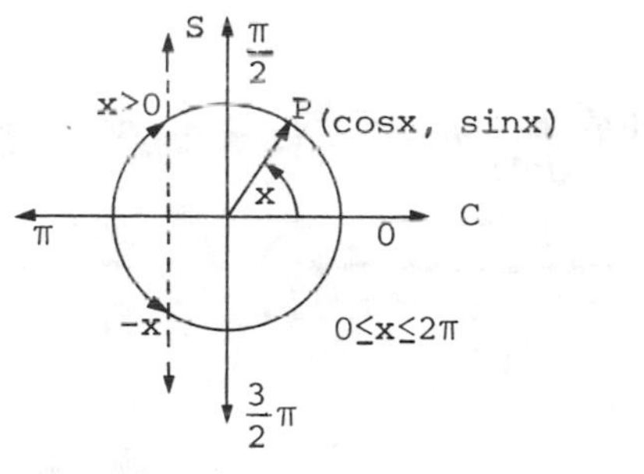

Fig. 3.1

If we let P(x) = (cos x, sin x), then for any x such that $0 \leq x \leq 2\pi$, the points P(x) and P(-x) lie on the same vertical axis. They are symmetrically located with respect to the c-axis.

This implies that cos(-x) = cos x and sin(-x) = -sin x

Another fundamental identity to remember is

$$\cos^2 x + \sin^2 x = 1$$

For any $x \in R$, $-1 \leq \cos x \leq 1$ and $-1 \leq \sin x \leq 1$, therefore, both the sine and cosine functions are continuous for all real numbers.

If PQR is an angle t and P has coordinates (x,y) on the unit circle, then by joining PR we get angle PRQ = 90° (Fig. 3.2), and then we can define all the trigonometric functions in the following way:

sine of t, $\sin t = y$

cosine of t, $\cos t = x$

tangent of t, $\tan t = \frac{y}{x}$, $x \neq 0$

cotangent of t, $\tan t = \frac{x}{y}$, $y \neq 0$

secant of t, $\sec t = \frac{1}{x}$, $x \neq 0$

cosecant of t, $\csc t = \frac{1}{y}$, $y \neq 0$.

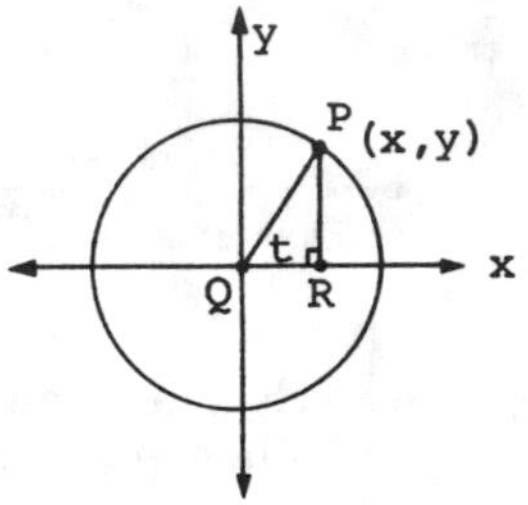

Fig. 3.2

Provided the denominators are not zero, the following relationships exist:

$$\sin t = \frac{1}{\csc t} \qquad \tan t = \frac{\sin t}{\cos t}$$

$$\cos t = \frac{1}{\sec t} \qquad \cot t = \frac{\cos t}{\sin t}$$

$$\tan t = \frac{1}{\cot t}$$

More Trigonometric Identities can be found in the appendix of this book.

Figures 3.3 and 3.4 show the graphs of each of the trigonometric functions. Notice that the x-axis is measured in radians.

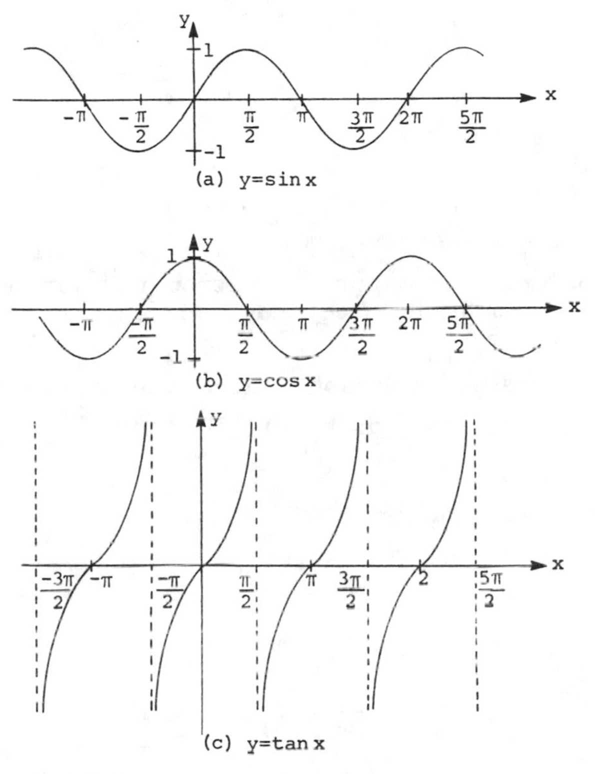

(a) y=sin x

(b) y=cos x

(c) y=tan x

Fig. 3.3

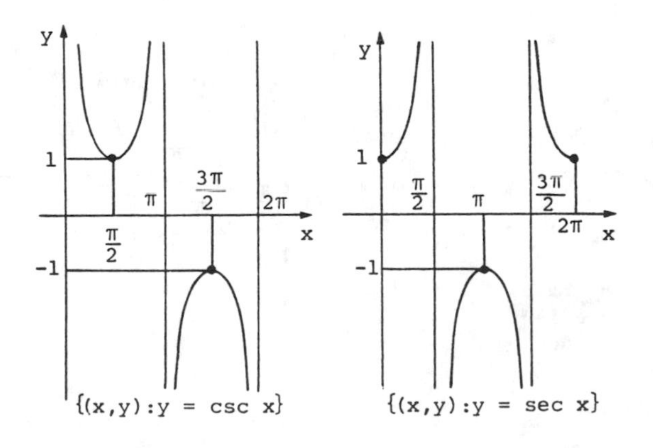

{(x,y):y = csc x}

{(x,y):y = sec x}

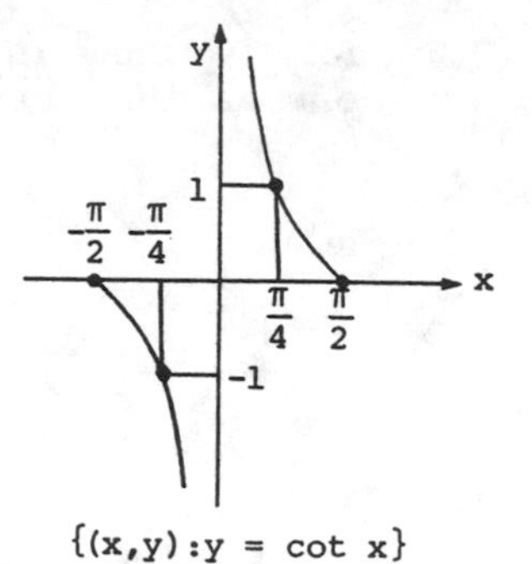

$\{(x,y): y = \cot x\}$

Fig. 3.4

In order to graph a trigonometric function we must know the amplitude, frequency, phase angle and the period of the function.

For example, to graph a function of the form

$$y = a \sin(bx+c)$$

we must determine:

$$a = \text{amplitude}$$

$$b = \text{frequency}$$

$$\frac{c}{b} = \text{phase angle}$$

and $$\frac{2\pi}{b} = \text{period.}$$

Let us graph the function $y = 2 \sin(2x + \frac{\pi}{4})$. Amplitude = 2, period = $\frac{2\pi}{2} = \pi$, phase $\measuredangle = \frac{\pi}{8}$.

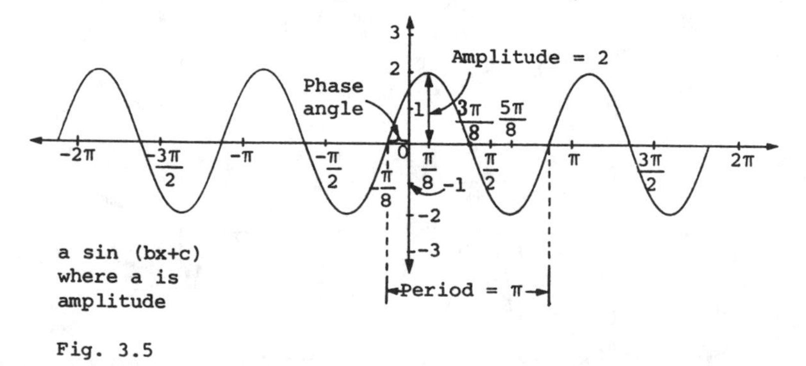

a sin (bx+c)
where a is
amplitude

Fig. 3.5

3.2 INVERSE TRIGONOMETRIC FUNCTIONS

By restricting the domain of the trigonometric functions we can define their inverse functions.

The inverse sine function, denoted $\sin^{-1}$, is defined to be $\sin^{-1}x = y$ if and only if $\sin y = x$ where $-1 \leq x \leq 1$ and $-\pi/2 \leq y \leq \frac{\pi}{2}$.

In a similar manner we define:

$-\frac{\pi}{2} \leq \sin^{-1}x \leq \frac{\pi}{2}$	$-1 \leq x \leq 1$	monotone increasing
$0 \leq \cos^{-1}x \leq \pi$	$-1 \leq x \leq 1$	monotone decreasing
$\frac{-\pi}{2} < \tan^{-1}x < \frac{\pi}{2}$	for all $x \in R$	
$0 < \cot^{-1}x < \pi$	for all $x \in R$	
$0 \leq \sec^{-1}x \leq \pi$	$\lvert x \rvert > 1$	
$-\frac{\pi}{2} \leq \csc^{-1}x \leq \frac{\pi}{2}$	$\lvert x \rvert \geq 1$	

3.3 EXPONENTIAL AND LOGARITHMIC FUNCTIONS

If f is a nonconstant function that is continuous and satisfies the functional equation $f(x+y) = f(x) \cdot f(y)$, then $f(x) = a^x$ for some constant a. That is, f is an exponential function.

Consider the exponential function a^x, $a > 0$ and the logarithmic function $\log_a x$, $a > 0$. Then a^x is defined for all $x \in R$, and $\log_a x$ is defined only for positive $x \in R$.

These functions are inverses of each other,

$$a^{\log_a x} = x; \log_a(a^y) = y.$$

Let a^x, $a > 0$ be an exponential function. Then for any real numbers x and y

a) $a^x \cdot a^y = a^{x+y}$

b) $(a^x)^y = a^{xy}$

Let $\log_a x$, $a > 0$ be a logarithmic function. Then for any positive real numbers x and y

a) $\log_a(xy) = \log_a(x) + \log_a(y)$

b) $\log_a(x^y) = y \log_a(x)$

Let $h > -1$ be any real number. Then for any natural number $n \in N$,

$$(1+h)^n \geq 1 + nh.$$

3.3.1 THE NATURAL LOGARITHMIC FUNCTION

A) To every real number y there corresponds a unique positive real number x such that the natural logarithm, ln, of x is equal to y. That is $\ln x = y$.

B) The natural exponential function, denoted by exp, is defined by

$$\exp x = y \text{ if and only if } \ln y = x$$

for all x, where $y > 0$.

C) The natural log and natural exponential are inverse functions. $\ln(\exp x) = x$ and $\exp(\ln y) = y$.

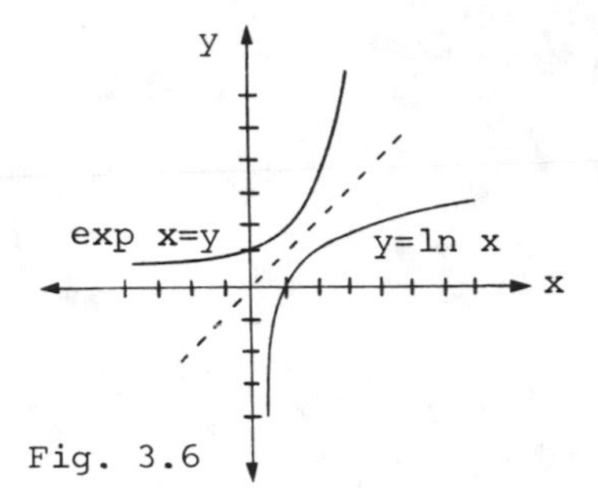

Fig. 3.6

D) The letter e denotes the unique positive real number such that ln e = 1.

E) If x is a real number then e^x is the unique real number y such that

$$e^x = y \text{ if and only if } \ln y = x.$$

F) If p and q are real numbers and r is rational then

a) $e^p e^q = e^{p+q}$

b) $\dfrac{e^p}{e^q} = e^{p-q}$

c) $(e^p)^r = e^{pr}$

CHAPTER 4

LIMITS

4.1 DEFINITION

Let f be a function that is defined on an open interval containing a, but possibly not defined at a itself. Let L be a real number. The statement

$$\lim_{x \to a} f(x) = L$$

defines the limit of the function f(x) at the point a. Very simply, L is the value that the function has as the point a is approached.

4.2 THEOREMS ON LIMITS

The following are important properties of limits:

Consider $\lim_{x \to a} f(x) = L$ and $\lim_{x \to a} g(x) = K$, then

A) Uniqueness – If $\lim_{x \to a} f(x)$ exists then it is unique.

B) $\lim_{x \to a} [f(x)+g(x)] = \lim_{x \to a} f(x) + \lim_{x \to a} g(x) = L+K$

C) $\lim_{x \to a} [f(x)-g(x)] = \lim_{x \to a} f(x) - \lim_{x \to a} g(x) = L-K$

D) $\lim_{x \to a} [f(x) \cdot g(x)] = \lim_{x \to a} f(x) \cdot \lim_{x \to a} g(x) = L \cdot K$

E) $\lim_{x \to a} \frac{f(x)}{g(x)} = \frac{\lim_{x \to a} f(x)}{\lim_{x \to a} g(x)} = \frac{L}{K}$ provided $K \neq 0$

F) $\lim_{x \to a} \frac{1}{g(x)} = \frac{1}{K}$, $K \neq 0$

G) $\lim_{x \to a} [f(x)]^n = [\lim_{x \to a} f(x)]^n$ for $n > 0$

H) $\lim_{x \to a} [cf(x)] = c[\lim_{x \to a} f(x)]$, $c \in R$

I) $\lim_{x \to a} cx^n = c \lim_{x \to a} x^n = ca^n$, $c \in R$

J) If f is a polynomial function then $\lim_{x \to a} f(x) = f(a)$ for all $a \in R$.

K) $\lim_{x \to a} \sqrt[n]{x} = \sqrt[n]{a}$ when $a \geq 0$ and n is a positive integer or when $a \leq 0$ and n is an odd positive integer.

L) $\lim_{x \to a} \sqrt[n]{f(x)} = \sqrt[n]{\lim_{x \to a} f(x)}$ when n is a positive integer

M) If $f(x) \leq h(x) \leq g(x)$ for all x in an open interval containing a, except possibly at a, and if $\lim_{x \to a} f(x) = L = \lim_{x \to a} g(x)$ then $\lim_{x \to a} h(x) = L$.

4.3 ONE-SIDED LIMITS

Suppose f is a function such that it is not defined for all values of x. Rather, it is defined in such a way that it "jumps" from one y value to the next instead of smoothly going from one y value to the next. Examples are shown in Fig. 4.1 and 4.2.

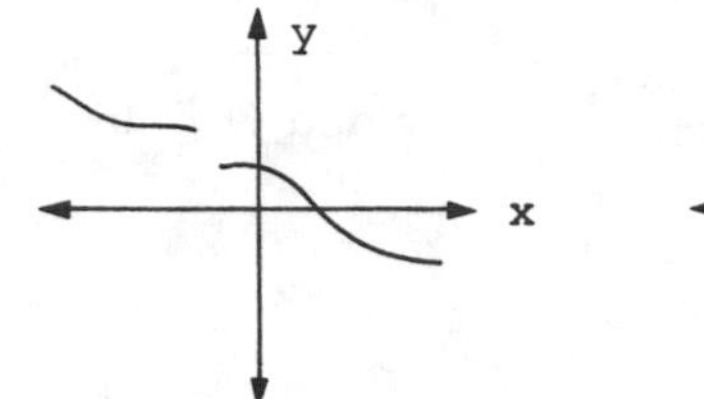

Fig. 4.1 y=f(x) is not defined for all x values.

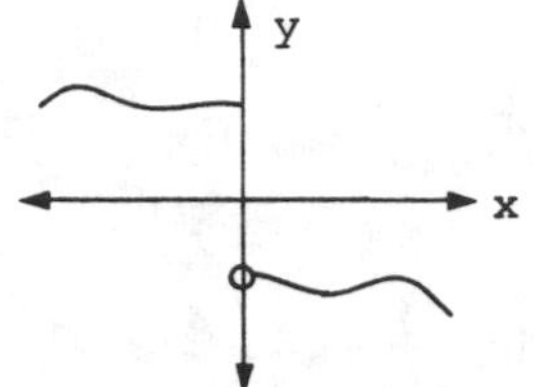

Fig. 4.2 y=f(x) "jumps" from a positive value to a negative one.

The statement $\lim_{x \to a^+} f(x) = R$ tells us that as x approaches "a" from the right or from positive infinity, the function f has the limit R.

Similarly, the statement $\lim_{x \to a^-} f(x) = L$ says that as x approaches "a" from the left-hand side or from negative infinity, the function f has the limit L.

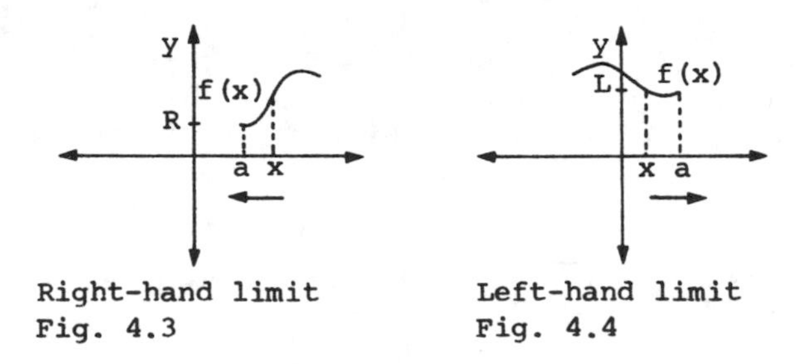

Right-hand limit
Fig. 4.3

Left-hand limit
Fig. 4.4

If f is defined in an open interval containing a, except possibly at a, then

$$\lim_{x \to a} f(x) = L \quad \text{if and only if}$$

$$\lim_{x \to a^+} f(x) = L = \lim_{x \to a^-} f(x).$$

Notice that in Fig. 4.2 the right-hand limit is not the same of the left-hand limit as it is in Fig. 4.5.

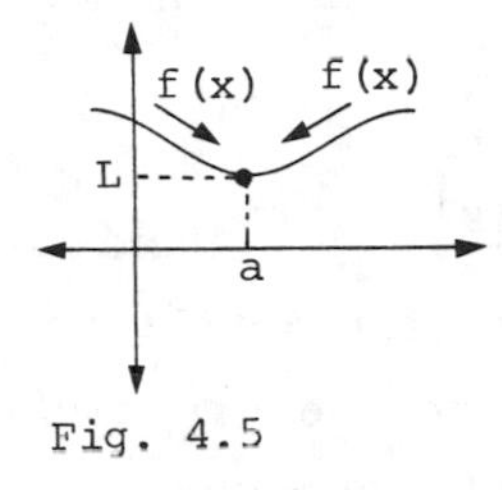

Fig. 4.5

4.4 SPECIAL LIMITS

A) $\lim_{x \to 0} \frac{\sin x}{x} = 1,\qquad \lim_{x \to 0} \frac{1-\cos x}{x} = 0$

B) $\lim_{n \to \infty} (1 + \frac{1}{n})^n = e,\qquad \lim_{n \to 0} (1+n)^{1/n} = e$

C) For $a > 1$ $\lim_{x \to +\infty} a^x = +\infty, \lim_{x \to -\infty} a^x = 0$

$$\lim_{x \to +\infty} \log_a x = +\infty,\ \lim_{x \to 0} \log_a x = -\infty$$

D) For $0 < a < 1$, $\lim_{x \to +\infty} a^x = 0$, $\lim_{x \to -\infty} a^x = +\infty$

$$\lim_{x \to +\infty} \log_a x = -\infty,\ \lim_{x \to 0} \log_a x = +\infty$$

Some nonexistent limits which are frequently encountered are:

A) $\lim_{x \to 0} \frac{1}{x^2}$, as x approaches zero, x^2 gets very small and also becomes zero therefore $\frac{1}{0}$ is undefined and the limit does not exist.

B) $\lim_{x \to 0} \frac{|x|}{x}$ does not exist.

Proof:

If $x > 0$, then $\frac{|x|}{x} = \frac{x}{x} = 1$ and hence lies to the right of the y-axis, the graph of f coincides with the line $y = 1$. If $x < 0$ then $\frac{-x}{x} = -1$ and the graph of f coincides with the line $y = -1$ to the left of the y-axis.

If it were true that $\lim_{x \to 0} \frac{|x|}{x} = L$ for some L, then the preceding remarks imply that $-1 \leq L \leq 1$.

If we consider any pair of horizontal lines $y = L \pm \varepsilon$, where $0 < \varepsilon < 1$, then there exists points on the graph which are not between these lines for some non-zero x in every interval $(-\delta, \delta)$ containing 0. It follows that the limit does not exist.

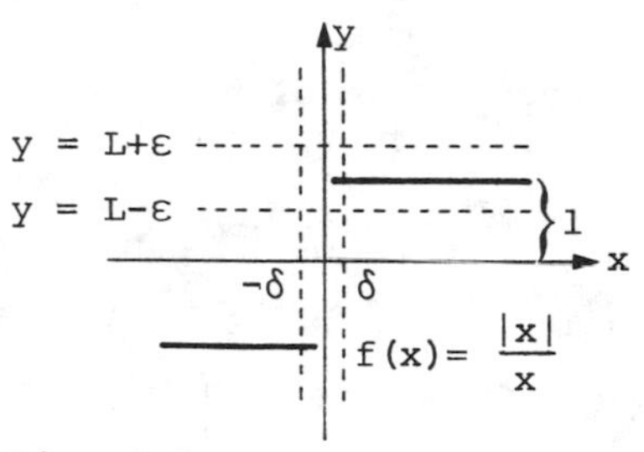

Fig. 4.6

4.5 CONTINUITY

A function f is continuous at a point a if

$$\lim_{x \to a} f(x) = f(a).$$

This implies that three conditions are satisfied:

a) $f(a)$ exists, that is, f is defined at a
b) $\lim_{x \to a} f(x)$ exists, and
c) the two numbers are equal.

To test continuity at a point $x = a$ we test whether

$$\lim_{x \to a^+} M(x) = \lim_{x \to a^-} M(x) = M(a)$$

4.5.1 THEOREMS ON CONTINUITY

A) A function defined in a closed interval $[a,b]$ is continuous in $[a,b]$ if and only if it is continuous in the open interval (a,b), as well as continuous from the right at "a" and from the left at "b".

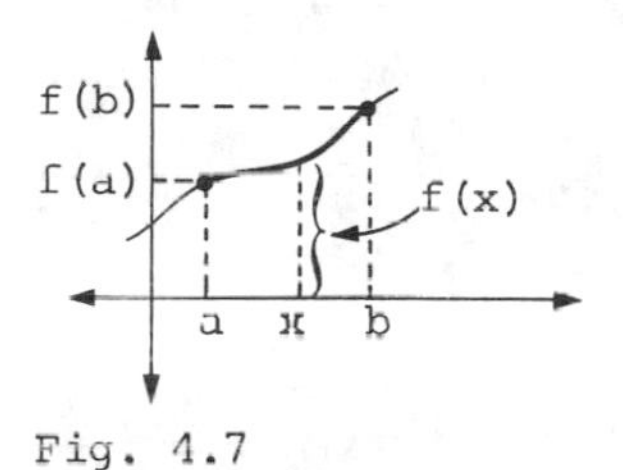

Fig. 4.7

B) If f and g are continuous functions at a, then so are the functions $f+g$, $f-g$, fg and f/g where $g(a) \neq 0$.

C) If $\lim_{x \to a} g(x) = b$ and f is continuous at b,

$$\text{then } \lim_{x \to a} f(g(x)) = f(b) = f[\lim_{x \to a} g(x)].$$

D) If g is continuous at a and f is continuous at $b = g(a)$, then

$$\lim_{x \to a} f(g(x)) = f[\lim_{x \to a} g(x)] = f(g(a)).$$

E) Intermediate Value Theorem. If f is continuous on a closed interval $[a,b]$ and if $f(a) \neq f(b)$, then f takes on every value between $f(a)$ and $f(b)$ in the interval $[a,b]$.

F) $f(x) = k$, $k \in R$ is continuous everywhere.

G) $f(x) = x$, the identity function is continuous everywhere.

H) If f is continuous at a, then $\lim_{n \to \infty} f(a + \frac{1}{n}) = f(a)$.

I) If f is continuous on an interval containing a and b, $a < b$, and if $f(a) \cdot f(b) < 0$ then there exists at least one point c, $a < c < b$ such that $f(c) = 0$.

CHAPTER 5

THE DERIVATIVE

5.1 THE DEFINITION AND Δ-METHOD

The derivative of a function expresses its rate of change with respect to an independent variable. The derivative is also the slope of the tangent line to the curve.

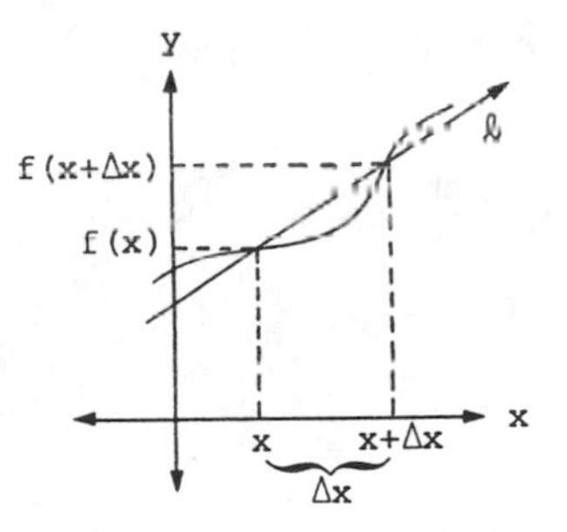

Fig. 5.1

Consider the graph of the function f in Fig. 5.1. Choosing a point x and a point $x + \Delta x$ (where Δx denotes a small distance on the x-axis) we can obtain both, f(x) and $f(x+\Delta x)$. Drawing a tangent line, ℓ, of the curve through the points f(x) and $f(x+\Delta x)$, we can measure the rate of change of this line. As we let the distance, Δx, approach zero, then

$$\lim_{\Delta x \to 0} \frac{f(x+\Delta x)-f(x)}{\Delta x}$$

becomes the instantaneous rate of change of the function or the derivative.

We denote the derivative of the function f to be f'. So we have

$$f'(x) = \lim_{\Delta x \to 0} \frac{f(x+\Delta x)-f(x)}{\Delta x}$$

If y = f(x), some common notations for the derivative are

$$y' = f'(x)$$

$$\frac{dy}{dx} = f'(x)$$

$$D_x y = f'(x) \text{ or } Df = f'$$

5.1.1 THE DERIVATIVE AT A POINT

If f is defined on an open interval containing "a", then

$$f'(a) = \lim_{x \to a} \frac{f(x)-f(a)}{x-a},$$

provided the limit exists.

5.2 RULES FOR FINDING THE DERIVATIVES

General rule:

A) If f is a constant function, f(x) = c, then f'(x) = 0.

B) If f(x) = x, then f'(x) = 1.

C) If f is differentiable, then $(cf(x))' = cf'(x)$

D) Power Rule If $f(x) = x^n$, $n \in Z$, then

$f'(x) = nx^{n-1}$; if $n < 0$ then x^n is not defined at $x = 0$.

E) If f and g are differentiable on the interval (a,b) then:

a) $(f+g)'(x) = f'(x) + g'(x)$

b) Product Rule. $(fg)'(x) = f(x)g'(x) + g(x)f'(x)$

Example: Find $f'(x)$ if $f(x) = (x^3+1)(2x^2+8x-5)$.

$$f'(x) = (x^3+1)(4x+8)+(2x^2+8x-5)(3x^2)$$

$$= 4x^4 + 8x^3 + 4x + 8 + 6x^4 + 24x^3 - 15x^2$$

$$= 10x^4 + 32x^3 - 15x^2 + 4x + 8$$

c) Quotient Rule: $\left(\frac{f}{g}\right)(x) = \frac{g(x)f'(x)-f(x)g'(x)}{[g(x)]^2}$

Example: Find $f'(x)$ if $f(x) = \frac{3x^2-x+2}{4x^2+5}$

$$f'(x) = \frac{-(3x^2-x+2)(8x)+(4x^2+5)(6x-1)}{(4x^2+5)^2}$$

$$= \frac{-(24x^3-8x^2+16x)+(24x^3-4x^2+30x-5)}{(4x^2+5)^2}$$

$$= \frac{4x^2+14x-5}{(4x^2+5)^2}$$

F) If $f(x) = x^{m/n}$, then $f'(x) = \frac{m}{n} x^{\frac{m}{n}-1}$

where $m, n \in Z$ and $n \neq 0$

G) Polynomials. If $f(x) = (a_0+a_1x+a_2x^2+...+a_nx^n)$

then $f'(x) = a_1+2a_2x+3a_3x^2+\ldots+na_nx^{n-1}$

This employs the power rule and rules concerning constants.

H) Chain Rule. Let f(u) be a composite function, where u=g(x).

Then $f'(u) = f'(u)g'(x)$ or if $y=f(u)$ and $u=g(x)$ then $D_xy = (D_uy)(D_xu) = f'(u)g'(x)$

5.3 IMPLICIT DIFFERENTIATION

An implicit function of x and y is a function in which one of the variables is not directly expressed in terms of the other. If these variables are not easily or practically separable, we can still differentiate the expression.

Apply the normal rules of differentiation such as the product rule, the power rule, etc. Remember also the chain rule which states $\frac{du}{dx} \times \frac{dx}{dt} = \frac{du}{dt}$.

Once the rules have been properly applied we will be left with as in the example of x and y, some factors of $\frac{dy}{dx}$.

We can then algebraically solve for the derivative $\frac{dy}{dx}$ and obtain the desired result.

5.4 TRIGONOMETRIC DIFFERENTIATION

The three most basic trigonometric derivatives are:

$$\frac{d}{dx}(\sin x) = \cos x,$$

$$\frac{d}{dx}(\cos x) = -\sin x,$$

$$\frac{d}{dx}(\tan x) = \sec^2 x.$$

Given any trigonometric function, it can be differentiated by applying these basics in combination with the general rules for differentiating algebraic expressions.

The following will be most useful if committed to memory:

$D_x \sin u = \cos u\, D_x u$

$D_x \cos u = -\sin u\, D_x u$

$D_x \tan u = \sec^2 u\, D_x u$

$D_x \sec u = \tan u \sec u\, D_x u$

$D_x \cot u = -\csc^2 u\, D_x u$

$D_x \csc u = -\csc u \cot u\, D_x u$

5.5 INVERSE TRIGONOMETRIC DIFFERENTIATION

Inverse trigonometric functions may be sometimes handled by inverting the expression and applying the rules for the direct trigonometric functions.

For example: $y = \sin^{-1} x$

$$D_x y = D_x \sin^{-1} x$$

$$= \frac{1}{\cos y} = \frac{1}{\sqrt{1-x^2}}, \quad |x| < 1.$$

Here are the derivatives for the inverse trigonometric functions which can be found in a manner similar to the above function:

$$D_x \sin^{-1}u = \frac{1}{\sqrt{1-u^2}} D_x u \quad , \quad |u| < 1$$

$$D_x \cos^{-1}u = \frac{-1}{\sqrt{1-u^2}} D_x u \quad , \quad |u| < 1$$

$$D_x \tan^{-1}u = \frac{1}{1+u^2} D_x u \quad , \quad \text{where } u = f(x) \text{ differentiable}$$

$$D_x \sec^{-1}u = \frac{1}{|u|\sqrt{u^2-1}} D_x u \ , \ u = f(x), \ |f(x)| > 1$$

$$D_x \cot^{-1}u = \frac{-1}{1+u^2} D_x u \quad , \quad u = f(x) \text{ differentiable}$$

$$D_x \csc^{-1}u = \frac{-1}{|u|\sqrt{u^2-1}} D_x u , u = f(x), \ |f(x)| > 1$$

5.6 EXPONENTIAL AND LOGARITHMIC DIFFERENTIATION

The exponential function e^x has the simplest of all derivatives. Its derivative is itself.

$$\frac{d}{dx} e^x = e^x$$

and

$$\frac{d}{dx} e^u = e^u \frac{du}{dx}$$

Since the natural logarithmic function is the inverse of $y = e^x$ and $\ln e = 1$, it follows that

$$\frac{d}{dx} \ln y = \frac{1}{y} \frac{dy}{dx}$$

and

$$\frac{d}{dx} \ln u = \frac{1}{u} \frac{du}{dx}$$

If x is any real number and a is any positive real number, then

$$a^x = e^{x \ln a}$$

From this definition we can obtain the following:

a) $\frac{d}{dx} a^x = a^x \ln a$ and $\frac{d}{dx} a^u = a^u \ln a \frac{du}{dx}$

b) $\frac{d}{dx} (\log_a x) = \frac{1}{x \ln a}$ and $\frac{d}{dx} \log_a |u| = \frac{1}{u \ln a} \frac{du}{dx}$
where $u \neq 0$

Sometimes it is useful to take the logs of a function and then differentiate since the computation becomes easier (as in the case of a product).

5.6.1 STEPS IN LOGARITHMIC DIFFERENTIATION

1. $y = f(x)$ — given
2. $\ln y = \ln f(x)$ — take logs and simplify
3. $D_x(\ln y) = D_x(\ln f(x))$ — differentiate implicitly
4. $\frac{1}{y} D_x y = D_x(\ln f(x))$
5. $D_x y = f(x) D_x(\ln f(x))$ — multiply by $y = f(x)$

To complete the solution it is necessary to differentiate $\ln f(x)$. If $f(x) < 0$ for some x then step 2 is invalid and we should replace step 1 by $|y| = |f(x)|$, and then proceed.

Example: $y = (x+5)(x^4+1)$

$$\ln y = \ln[(x+5)(x^4+1)] = \ln(x+5) + \ln(x^4+1)$$

$$\frac{d}{dx} \ln y = \frac{d}{dx} \ln(x+5) + \frac{d}{dx} \ln(x^4+1)$$

$$\frac{1}{y}\frac{dy}{dx} = \frac{1}{x+5} + \frac{4x^3}{x^4+1}$$

$$\frac{dy}{dx} = (x+5)(x^4+1) \left[\frac{1}{x+5} + \frac{4x^3}{x^4+1}\right]$$

$$= (x^4+1) + 4x^3(x+5)$$

This is the same result as obtained by using the product rule.

5.7 HIGH ORDER DERIVATIVES

The derivative of any function is also a legitimate function which we can differentiate. The second derivative can be obtained by:

$$\frac{d}{dx}\left[\frac{d}{dx}u\right] = \frac{d^2}{dx^2}u = u'' = D^2u,$$

where $u = g(x)$ is differentiable.

The general formula for higher orders and the nth derivative of u is,

$$\underbrace{\frac{d}{dx}\,\frac{d}{dx}\cdots\frac{d}{dx}}_{n \text{ times}}u = \frac{d^{(n)}}{dx^n}u = u^{(n)} = D_x^{(n)}u.$$

The rules for first order derivatives apply at each stage of higher order differentiation (e.g., sums, products, chain rule).

A function which satisfies the condition that its nth derivative is zero, is the general polynomial

$$p_{n-1}(x) = a_{n-1}x^{n-1} + a_{n-2}x^{n-2} + \ldots + a_0.$$

CHAPTER 6

APPLICATIONS OF THE DERIVATIVE

6.1 ROLLE'S THEOREM

Let f be continuous on a closed interval [a,b]. Assume f'(x) exists at each point in the open interval (a,b).

If f(a) = f(b) = 0 then there is at least one point (x_0) in (a,b) such that $f'(x_0) = 0$.

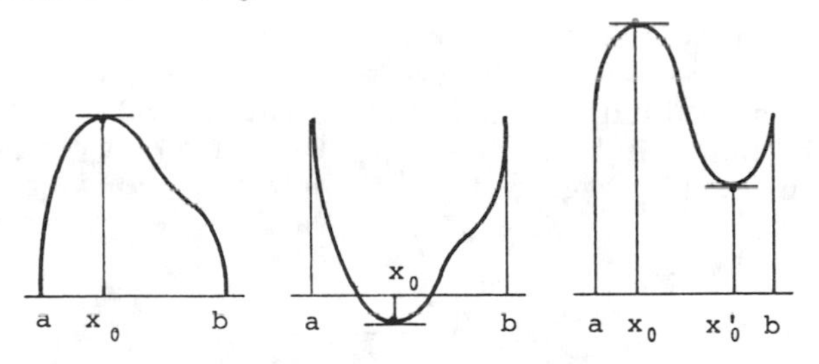

Fig. 6.1 Three functions which satisfy the hypotheses, hence the conclusion, of Rolle's theorem.

6.2 THE MEAN VALUE THEOREM

If f is continuous on [a,b] and has a derivative at every point in the interval (a,b), then there is at least one number c in (a,b) such that

$$f'(c) = \frac{f(b)-f(a)}{b-a}$$

Notice in Fig. 6.2 that the secant has slope

$$\frac{f(b)-f(a)}{b-a}$$

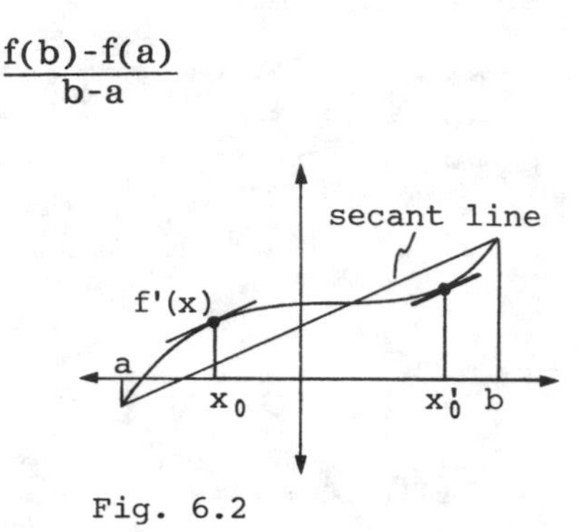

Fig. 6.2

and $f'(x)$ has slope of the tangent to the point $(x,f(x))$. For some x_0 in (a,b) these slopes are equal.

6.2.1 CONSEQUENCES OF THE MEAN VALUE THEOREM

A) If f is defined on an interval (a,b) and if $f'(x) = 0$ for each point in the interval, then $f(x)$ is constant over the interval. Fig. 6.3.

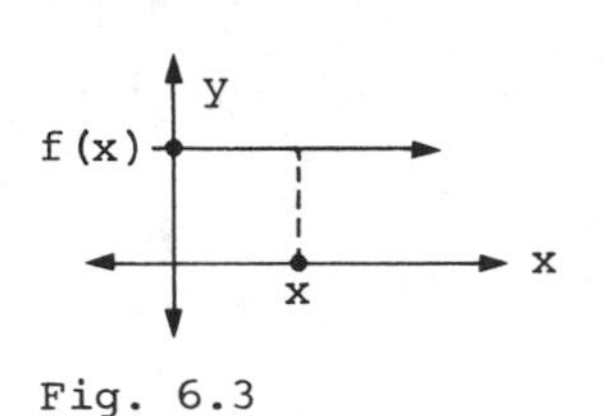

Fig. 6.3

B) Let f and g be differentiable on an interval (a,b). If, for each point x in the interval, $f'(x)$ and $g'(x)$ are equal, then there is a constant, c, such that

$$f(x) + c = g(x) \quad \text{for all } x.$$

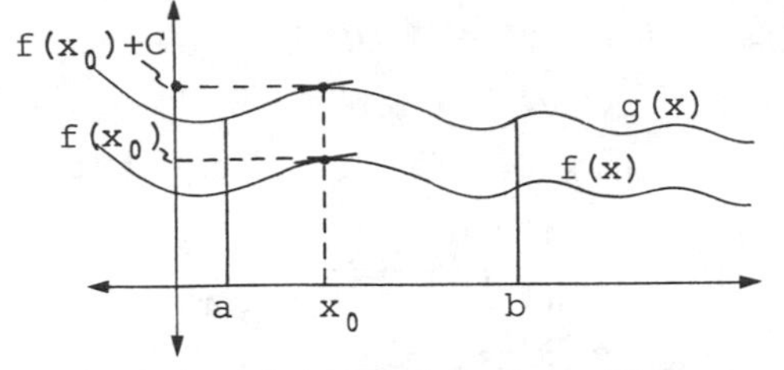

Fig. 6.4 $f(x)+C=g(x)$ for all x

C) The Extended Mean Value Theorem. Assume that the function f and its derivative f' are continuous on [a,b] and that f'' exists at each point x in (a,b), then there exists at least one point x_0, $a < x_0 < b$, such that

$$f(b) = f(a) + (b-a)f'(a) + \tfrac{1}{2}(b-a)^2 f''(x_0).$$

6.3 L'HÔPITAL'S RULE

An application of the Mean Value Theorem is in the evaluation of

$$\lim_{x \to a} \frac{f(x)}{g(x)} \text{ where } f(a)=0 \text{ and } g(a)=0.$$

L'Hôpital's Rule states that if the $\lim_{x \to a} \frac{f(x)}{g(x)}$ is an indeterminate form (i.e., $\frac{0}{0}$ or $\frac{\infty}{\infty}$), then we can differentiate the numerator and the denominator separately and arrive at an expression that has the same limit as the original problem.

Thus, $$\lim_{x \to a} \frac{f(x)}{g(x)} = \lim_{x \to a} \frac{f'(x)}{g'(x)}$$

In general, if f(x) and g(x) have properties

1) $f(a) = g(a) = 0$

2) $f^{(k)}(a) = g^{(k)}(a) = 0$ for $k=1,2,\ldots n$

but 3) $f^{(n+1)}(a)$ or $g^{(n+1)}(a)$ is not equal to zero, then

$$\lim_{x \to a} \frac{f(x)}{g(x)} = \frac{f^{(n+1)}(x)}{g^{(n+1)}(x)}$$

6.4 TANGENTS AND NORMALS

6.4.1 TANGENTS

A line which is tangent to a curve at a point "a", must have the same slope as the curve. That is, the slope of the tangent is simply

$$m = \lim_{h \to 0} \frac{f(a+h)-f(a)}{h}$$

Therefore, if we find the derivative of a curve and evaluate for a specific point, we obtain the slope of the curve and the tangent line to the curve at that point.

A curve is said to have a vertical tangent at a point $(a,f(a))$ if f is continuous at "a" and $\lim_{x \to a} |f'(x)| = \infty$.

6.4.2 NORMALS

A line normal to a curve at a point must have a slope perpendicular to the slope of the tangent line. If $f'(x) \neq 0$ then the equation for the normal line at a point (x_0, y_0) is

$$\boxed{y - y_0 = \frac{-1}{f'(x_0)}(x - x_0).}$$

6.5 MINIMUM AND MAXIMUM VALUES

If a function f is defined on an interval I, then

a) f is increasing on I if $f(x_1) < f(x_2)$ whenever x_1, x_2 are in I and $x_1 < x_2$.

b) f is decreasing on I if $f(x_1) > f(x_2)$ whenever $x_1 < x_2$ in I.

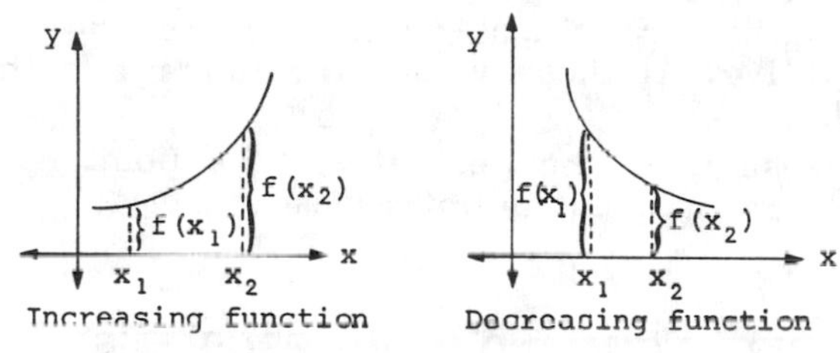

Increasing function Decreasing function

c) f is constant if $f(x_1) = f(x_2)$ for every x_1, x_2 in I.

Suppose f is defined on an open interval I and c is a number in I then,

a) f(c) is a local maximum value if $f(x) \leq f(c)$ for all x in I.

b) f(c) is a local minimum value if $f(x) \geq f(c)$ for all x in I.

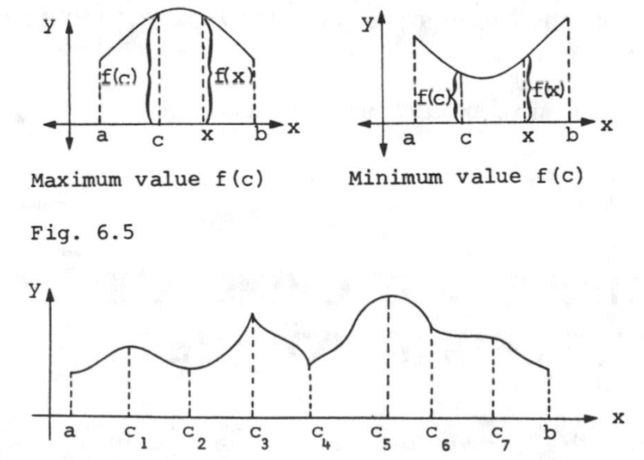

Maximum value f(c) Minimum value f(c)

Fig. 6.5

Fig. 6.6

In Fig. 6.6 in the interval [a,b], the local maxima occur at c_1, c_3, c_5 with an absolute maximum at c_5 and local minima occur at c_2 , c_4 .

To find Absolute Extrema for functions, first calculate f(c) for each critical number c, then calculate f(a) and f(b). The absolute extrema of f on [a,b] will then be the largest and the smallest of these functional values. If f(a) or f(b) is an extremum we call it an endpoint extremum.

Viewing the derivative as the slope of a curve, there may be points (or critical values) where the curve has a zero derivative. At these values the tangent to the curve is horizontal.

Conversely, if the derivative at a point exists and is not zero, then the point is not a local extrema.

6.5.1 SOLVING MAXIMA AND MINIMA PROBLEMS

Step 1. Determine which variable is to be maximized or minimized (i.e., the dependent variable y).

Step 2. Find the independent variable x.

Step 3. Write an equation involving x and y. All other variables can be eliminated by substitution.

Step 4. Differentiate with respect to the independent variable.

Step 5. Set the derivative equal to zero to obtain critical values.

Step 6. Determine maxima and minima.

6.6 CURVE SKETCHING AND THE DERIVATIVE TESTS

Using the knowledge we have about local extrema and the following properties of the first and second derivatives of a function, we can gain a better understanding of the graphs (and thereby the nature) of a given function.

A function is said to be smooth on an interval (a,b) if both f' and f'' exist for all $x \in (a,b)$.

6.6.1 THE FIRST DERIVATIVE TEST

Suppose that c is a critical value of a function, f, in an interval (a,b), then if f is continuous and differentiable we can say that,

a) if $f'(x) > 0$ for all $a < x < c$ and $f'(x) < 0$ for all $c < x < b$, then $f(c)$ is a local maximum.

b) if $f'(x) < 0$ for $a < x < c$ and $f'(x) > 0$ for $c < x < b$, then $f(c)$ is a local minimum.

c) if $f'(x) > 0$ or if $f'(x) < 0$ for all $x \in (a,b)$ then $f(c)$ is not a local extrema.

6.6.2 CONCAVITY

If a function is differentiable on an open interval containing c, then the graph at this point is

a) concave upward (or convex) if $f''(c) > 0$;

b) concave downward if $f''(c) < 0$.

If a function is concave upward than f' is increasing as x increases. If the function is concave downward, f' is decreasing as x increases.

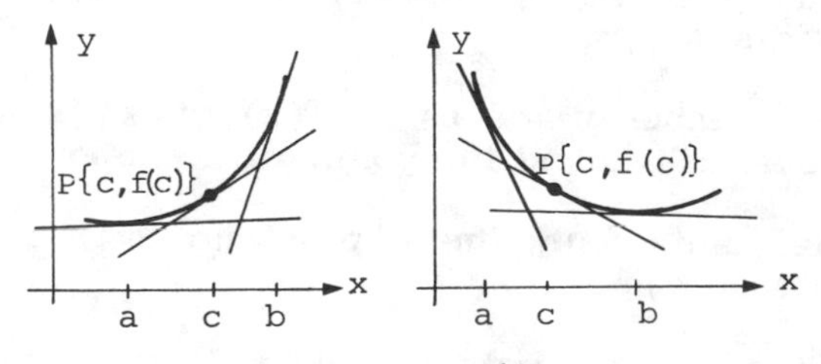

Upward concavity

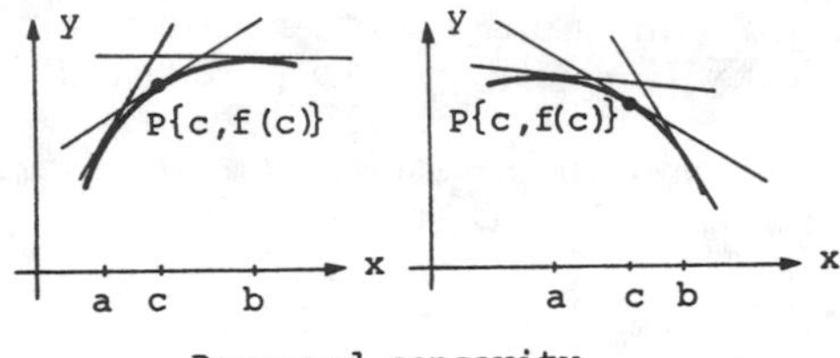

Downward concavity

Fig. 6.7

6.6.3 POINTS OF INFLECTION

Points which satisfy $f''(x) = 0$ may be positions where concavity changes. These points are called the points of inflection. It is the point at which the curve crosses its tangent line.

6.6.4 GRAPHING A FUNCTION USING THE DERIVATIVE TESTS

The following steps will help us gain a rapid understanding of a function's behavior.

A) Look for some basic properties such as oddness, evenness, periodicity, boundedness, etc.

B) Locate all the zeros by setting $f(x) = 0$.

C) Determine any singularities, $f(x) = \infty$.

D) Set $f'(x)$ equal to zero to find the critical values.

E) Find the points of inflection by setting $f''(x) = 0$.

F) Determine where the curve is concave, $f''(x) < 0$, and where it is convex $f''(x) > 0$.

G) Determine the limiting properties and approximations for large and small $|x|$.

H) Prepare a table of values x, $f(x)$, $f'(x)$ which includes the critical values and the points of inflection.

I) Plot the points found in Step H and draw short tangent lines at each point.

J) Draw the curve making use of the knowledge of concavity and continuity.

6.7 RECTILINEAR MOTION

When an object moves along a straight line we call the motion rectilinear motion. Distance s, velocity v, and acceleration a, are the chief concerns of the study of motion.

Velocity is the proportion of distance over time.

$$v = \frac{s}{t}$$

$$\text{Average velocity} = \frac{f(t_2)-f(t_1)}{t_2 - t_1}$$

where t_1, t_2 are time instances and $f(t_2)-f(t_1)$ is the displacement of an object.

Instantaneous velocity at time t is defined as

$$v = D\ s(t) = \lim_{h \to 0} \frac{f(t+h)-f(t)}{h}$$

We usually write

$$v(t) = \frac{ds}{dt}.$$

Acceleration, the rate of change of velocity with respect to time is

$$a(t) = \frac{dv}{dt}.$$

It follows clearly that

$$a(t) = v'(t) = s''(t).$$

When motion is due to gravitational effects, g = 32.2

ft/sec^2 or $g = 9.81$ m/sec^2 is usually substituted for acceleration.

Speed at time t is defined as $|v(t)|$. The speed indicates how fast an object is moving without specifying the direction of motion.

6.8 RATE OF CHANGE AND RELATED RATES

6.8.1 RATE OF CHANGE

In the last section we saw how functions of time can be expressed as velocity and acceleration. In general, we can speak about the rate of change of any function with respect to an arbitrary parameter (such as time in the previous section).

For linear functions $f(x) = mx+b$, the rate of change is simply the slope m.

For non-linear functions we define the

1) average rate of change between points c and d to be

$$\frac{f(d)-f(c)}{d-c}$$

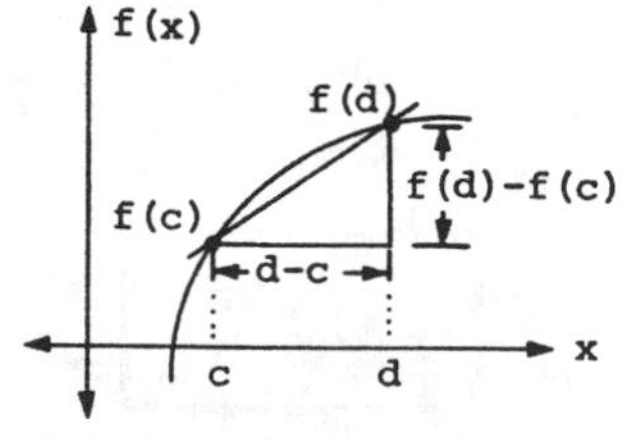

Fig. 6.8

2) instantaneous rate of change of f at the point x to be

$$f'(x) = \lim_{h \to 0} \frac{f(x+h)-f(x)}{h}$$

If the limit does not exist, then the rate of change of f at x is not defined.

The form, common to all related rate problems, is as follows:

a) Two variables, x and y are given. They are functions of time, but the explicit functions are not given.

b) The variables, x and y are related to each other by some equation such as $x^2+y^3-2x-7y^2+2 = 0$.

c) An equation which involves the rate of change $\frac{dx}{dt}$ and $\frac{dy}{dt}$ is obtained by differentiating with respect to t and using the chain rule.

As an illustration, the previous equation leads to

$$2x\frac{dx}{dt} + 3y^2\frac{dy}{dt} - 2\frac{dx}{dt} - 14y\frac{dy}{dt} = 0$$

The derivatives $\frac{dx}{dt}$ and $\frac{dy}{dt}$ in this equation are called the related rates.

CHAPTER 7

THE DEFINITE INTEGRAL

7.1 ANTIDERIVATIVES

Definition:

If $F(x)$ is a function whose derivative $F'(x) = f(x)$, then $F(x)$ is called the antiderivative of $f(x)$.

THEOREM:

If $F(x)$ and $G(x)$ are two antiderivatives of $f(x)$, then $F(x) = G(x) + c$, where c is a constant.

7.1.1 POWER RULE FOR ANTIDIFFERENTIATION

Let "a" be any real number, "r", any rational number not equal to -1 and "c" an arbitrary constant.

$$\text{If } f(x) = ax^r, \text{ then } F(x) = \frac{a}{r+1}x^{r+1} + c.$$

THEOREM:

An antiderivative of a sum is the sum of the antiderivatives.

$$\frac{d}{dx}(F_1+F_2) = \frac{d}{dx}(F_1) + \frac{d}{dx}(F_2) = f_1 + f_2$$

7.2 AREA

To find the area under the graph of a function f from a to b, we divide the interval [a,b] into n subintervals, all having the same length (b-a)/n. This is illustated in Figure 7.1.

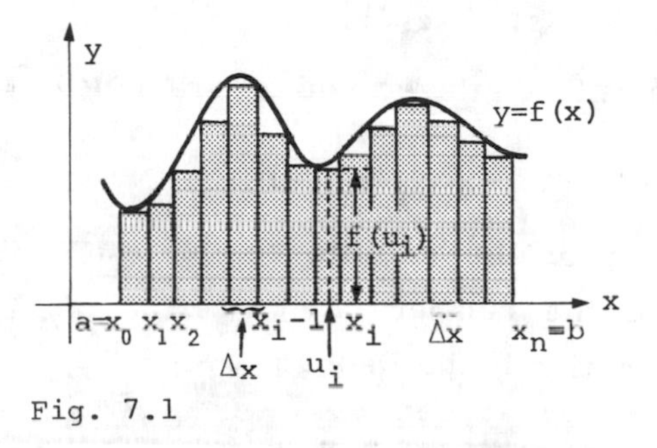

Fig. 7.1

Since f is continuous on each subinterval, f takes on a minimum value at some number u_i in each subinterval.

We can construct a rectangle with one side of length $[x_{i-1}, x_i]$, and the other side of length equal to the minimum distance $f(u_i)$ from the x-axis to the graph of f.

The area of this rectangle is $f(u_i)\Delta x$. The boundary of the region formed by the sum of these rectangles is called the inscribed rectangular polygon.

The area (A) under the graph of f from a to b is

$$A = \lim_{\Delta x \to 0} \sum_{i=1} f(u_i)\Delta x.$$

The area A under the graph may also be obtained by means of circumscribed rectangular polygons.

In the case of the circumscribed rectangular polygons the maximum value of f on the interval $[x_{i-1}, x_i]$, v_i, is used.

Note that the area obtained using circumscribed rectangular polygons should always be larger than that obtained using inscribed rectangular polygons.

7.3 DEFINITION OF DEFINITE INTEGRAL

A partition P of a closed interval [a,b] is any decomposition of [a,b] into subintervals of the form,

$[x_0,x_1], [x_1,x_2], [x_2,x_3],\ldots,[x_{n-1},x_n]$

where n is a positive integer and x_i are numbers, such that

$a = x_0<x_1<x_2<\ldots<x_{n-1}<x_n = b.$

The length of the subinterval is $\Delta x_i = x_i - x_{i-1}$. The largest of the numbers $\Delta x_1, \Delta x_2 \ldots \Delta x_n$ is called the norm of the partition P and denoted by $||P||$.

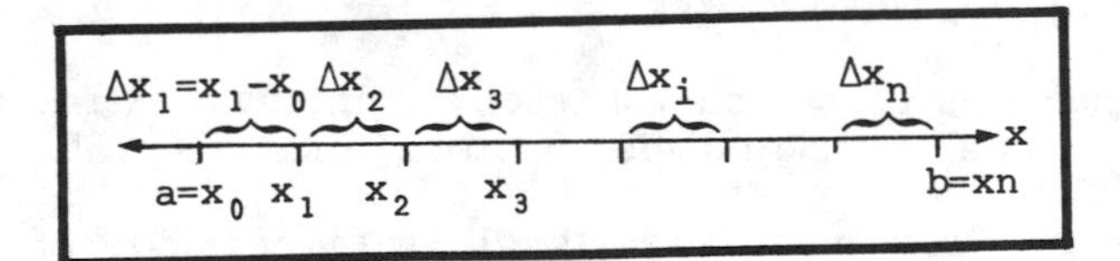

Definition:

Let f be a function that is defined on a closed interval [a,b] and let P be a partition of [a,b]. A Riemann Sum of f for P is any expression R_p of the form,

$$R_p = \sum_{i=1}^{n} f(w_i)\Delta x_i,$$

where w_i is some number in $[x_{i-1},x_i]$ for $i = 1,2,\ldots,n$.

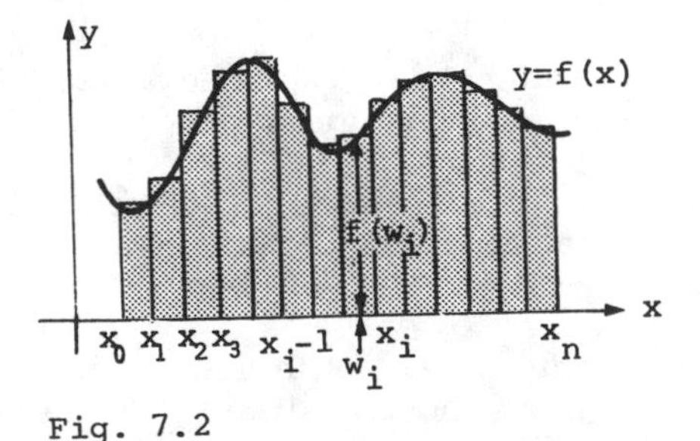

Fig. 7.2

Definition:

Let f be a function that is defined on a closed interval [a,b]. The definite integral of f from a to b, denoted by $\int_a^b f(x)dx$ is given by

$$\int_a^b f(x)dx = \lim_{||P|| \to 0} \sum_i f(w_i)\Delta x_i ,$$

provided the limit exists.

THEOREM :

If f is continuous on [a,b], then f is integrable on [a,b].

THEOREM :

If f(a) exists, then $\int_a^a f(x)dx = 0.$

7.4 PROPERTIES OF THE DEFINITE INTEGRAL

A) If f is integrable on [a,b], and k is any real number, then kf is integrable on [a,b] and

$$\boxed{\int_a^b kf(x)dx = k \int_a^b f(x)dx .}$$

B) If f and g are integrable on [a,b], then f+g is integrable on [a,b] and

$$\int_a^b [f(x)+g(x)]dx = \int_a^b f(x)dx + \int_a^b g(x)dx.$$

C) If $a<c<b$ and f is integrable on both [a,c] and [c,b] , then f is integable on [a,b] and

$$\int_a^b f(x)dx = \int_a^c f(x)dx + \int_c^b f(x)dx.$$

D) If f is integrable on a closed interval and if a, b, and c are any three numbers in the interval, then

$$\int_a^b f(x)dx = \int_a^c f(x)dx + \int_c^b f(x)dx.$$

E) If f is integrable on [a,b] and if $f(x) \geq 0$ for all x in [a,b], then $\int_a^b f(x)dx \geq 0$.

7.5 THE FUNDAMENTAL THEOREM OF CALCULUS

The fundamental theorem of calculus establishes the relationship between the indefinite integrals and differentiation by use of the mean value theorem.

7.5.1 MEAN VALUE THEOREM FOR INTEGRALS

If f is continuous on a closed interval [a,b], then

there is some number P in the open interval (a,b) such that

$$\int_a^b f(x)dx = f(P)(b-a)$$

To find f(P) we divide both sides of the equation by (b-a) obtaining

$$f(P) = \frac{1}{b-a} \int_a^b f(x)dx.$$

7.5.2 DEFINITION OF THE FUNDAMENTAL THEOREM

Suppose f is continuous on a closed interval [a,b], then

a) If the function G is defined by

$$G(x) = \int_a^x f(t)dt,$$

for all x in [a,b], then G is an antiderivative of f on [a,b].

b) If F is any antiderivative of f, then

$$\boxed{\int_a^b f(x)dx = F(b)-F(a)}$$

7.6 INDEFINITE INTEGRAL

The indefinite integral of f(x), denoted by $\int f(x)dx$, is the most general integral of f(x), that is

$$\int f(x)dx = F(x) + C.$$

F(x) is any function such that $F'(x) = f(x)$. C is an arbitrary constant.

7.6.1 INTEGRATION FORMULAS

Table 7.1

1. $\int x^n dx = \frac{x^{n+1}}{n+1} + C, \quad n \neq -1$	7. $\int \sin ax dx = -\frac{1}{a}\cos ax + C$
2. $\int \frac{dx}{x} = \ln \lvert x \rvert + C$	8. $\int \cos ax dx = \frac{1}{a}\sin ax + C$
3. $\int \frac{dx}{x-a} = \ln \lvert x-a \rvert + C$	9. $\int \sec^2 x\, dx = \tan x + C$
4. $\int \frac{dx}{x^2+a^2} = \frac{1}{a}\tan^{-1}\frac{x}{a} + C$	10. $\int e^{ax} dx = \frac{e^{ax}}{a} + C$
5. $\int \frac{x dx}{x^2+a^2} = \frac{1}{2}\ln \lvert x^2+a^2 \rvert + C$	11. $\int \sinh ax dx = \frac{1}{a}\cosh ax + C$
6. $\int \frac{dx}{(a^2-x^2)^{\frac{1}{2}}} = \sin^{-1}\frac{x}{a} + C$	12. $\int \cosh ax dx = \frac{1}{a}\sinh ax + C$

$\ln x \equiv \log_e x$ is called the logarithm of base e where $e \equiv 2.7182818$ ---

7.6.2 ALGEBRAIC SIMPLIFICATION

Certain apparently complicated integrals can be made simple by simple algebraic manipulations.

Example: Find $\int \frac{x}{x+1} dx$

Write $\frac{x}{x+1} = \frac{x+1-1}{x+1} = 1 - \frac{1}{x+1}$

$$\int \frac{x}{x+1} dx = \int dx - \int \frac{dx}{x+1} = x - \ln |x+1| + c$$

7.6.3 SUBSTITUTION OF VARIABLES

Suppose $F(x)$ is expressed as a composite function, $F(x) = f(u(x))$, then the differential $F'(x)\,dx = f'(u)du$.

Therefore, $\int F'(x)dx = \int f'(u)du = f(u)+c$

$$= f(u(x)) + c = F(x) + c.$$

THEOREM:

Let f and u be functions satisfying the following conditions:

a) f is continuous on a domain including the closed interval $\{x : a \leq x \leq b\}$.

b) For each point t in the closed interval $\{t : \alpha \leq t \leq \beta\}$, the value $u(t)$ is a point in $\{x : a \leq x \leq b\}$.

c) $u(\alpha) = a$, and $u(\beta) = b$.

d) u is continuous on $\{t : \alpha \leq t \leq \beta\}$.

The $\int_a^b f(x)dx = \int_\alpha^\beta f(u(t)) \cdot u'(t)dt.$

Example: Evaluate $\int \frac{x}{x^2+a^2}\,dx$

Let $u = x^2 + a^2$

$du = 2x\,dx$

$\frac{1}{2}du = x dx$

$$\int \frac{x}{x^2+a^2}\,dx = \frac{1}{2}\int \frac{du}{u} = \frac{1}{2}\,\mathrm{Ln}\,|u| + c$$

$$= \frac{1}{2}\,\mathrm{Ln}\,|x^2+a^2| + c$$

7.6.4 CHANGE OF VARIABLES

Example: Evalute $\int_0^1 x(1+x)^{\frac{1}{2}}dx$

Let $u = 1+x$, $du = dx$, $x = u-1$

$$\int_0^1 x(1+x)^{\frac{1}{2}} = \int_1^2 (u-1)u^{\frac{1}{2}}\,du.$$

*Notice the change in the limits for x=0, u=1 and for x=1 u=2.

$$\int_1^2 (u-1)u^{\frac{1}{2}}\,du = \int_1^2 u^{3/2} - u^{\frac{1}{2}}\,du$$

$$= 2/5\;u^{5/2} - 2/3\;u^{3/2}\Big|_1^2$$

$$=[(2/5)\sqrt{32} - (2/3)\sqrt{8})] - \left(\frac{2}{5} - \frac{2}{3}\right)$$

$$= \frac{4\sqrt{2}}{15} - \frac{4}{15} = \frac{4}{15}(\sqrt{2} - 1).$$

7.6.5 INTEGRATION BY PARTS

This method is based on the formula

$$d(uv) = u\,dv + v\,du.$$

The corresponding integration formula,

$uv = \int u\,dv + \int v\,du$, is applied in the form

$$\boxed{\int u\,dv = uv - \int v\,du}$$

This procedure involves the identification of u and dv and their manipulation into the form of the latter equation. v must be easily determined. If a definite integral is involved, then

$$\int_a^b u \frac{dv}{dx}\, dx = uv\Big]_a^b - \int_a^b v \frac{du}{dx}\, dx.$$

Example: Evaluate $\int x \cos x \, dx$

$u = x \quad dv = \cos x \, dx$

$du = dx \quad v = \sin x$

$$\int x \cos x \, dx = x \sin x - \int \sin x \, dx$$

$$= x \sin x - (-\cos x) + c$$

$$= x \sin x + \cos x + c$$

7.6.6 TRIGONOMETRIC INTEGRALS

Integrals of the form $\int \sin^n x \, dx$ or $\int \cos^n x dx$ can be evaluated without resorting to integration by parts. This is done in the following manner;

We write $\int \sin^n x \, dx = \int \sin^{n-1} \sin x \, dx$, if n is odd.

Since the integer n-1 is even, we may then use the fact that $\sin^2 x = 1-\cos^2 x$ to obtain a form which is easier to integrate.

Example: $\int \sin^5 x \, dx = \int \sin^4 x \sin x \, dx$

$$= \int (\sin^2 x)^2 \sin x \, dx$$

but $\sin^2 x = 1 - \cos^2 x$.

Hence, $\int \sin^5 x \, dx = \int (1-\cos^2 x)^2 \sin x \, dx$

$$= \int (1-2\cos^2 x+\cos^4 x)\sin x \, dx$$

Substitute $u = \cos x$, $du = -\sin x\, dx$

$$= -\int (1-2u^2+u^4)du = -u + \frac{2}{3}u^3 - \frac{u^5}{5}$$

$$= -\cos x + \frac{2}{3}\cos^3 x - \frac{1}{5}\cos^5 x + c.$$

A similar technique can be employed for odd powers of $\cos x$.

If the integrand is $\sin^n x$ or $\cos^n x$ and n is even, then the half angle formulas,

$$\sin^2 x = \frac{1 - \cos 2x}{2} \quad \text{or}$$

$$\cos^2 x = \frac{1+\cos 2x}{2}$$

may be used to simplify the integrand.

Example: $\int \cos^2 x\, dx = \frac{1}{2}\int (1+\cos 2x)dx$

$$= \frac{1}{2}x + \frac{1}{4}\sin 2x + c$$

CHAPTER 8

COMBINATIONS AND PERMUTATIONS

This chapter is a particularly helpful review of the material needed in the study of probability and statistics which uses calculus techniques. Statistics is a branch of science which is an outgrowth of the theory of probability. Combinations and permutations are used in both statistics and probability, and they, in turn, involve operations with factorial notation. Therefore, combinations, permutations, and factorial notation are discussed in this chapter.

DEFINITIONS

A combination is defined as a possible selection of a certain number of objects taken from a group with no regard given to order. For instance, suppose we were to choose two letters from a group of three letters. If the group of three letters were A, B, and C, we could choose the letters in combinations of two as follows:

AB, AC, BC

The order in which we wrote the letters is of no concern. That is, AB could be written BA but we would still have only one combination of the letters A and B.

If order were considered, we would refer to the letters as permutations and make a distinction between AB and BA. The permutations of two letters from the group of three letters would be as follows:

AB, AC, BC, BA, CA, CB

The symbol used to indicate the foregoing combination will be ${}_3C_2$, meaning a group of three objects taken two at a time. For the previous permutation we will use ${}_3P_2$, meaning a group of three objects taken two at a time and ordered.

An understanding of factorial notation is required prior to a detailed discussion of combinations and permutations. We define the product of the integers 1 through n as n factorial and use the symbol n! to denote this. That is,

$$3! = 1 \cdot 2 \cdot 3$$

$$6! = 1 \cdot 2 \cdot 3 \cdot 4 \cdot 5 \cdot 6$$

$$n! = 1 \cdot 2 \cdot 3 \cdots (n - 1) \cdot n$$

EXAMPLE: Find the value of 5!
SOLUTION: Write

$$5! = 1 \cdot 2 \cdot 3 \cdot 4 \cdot 5$$

$$= 120$$

EXAMPLE: Find the value of

$$\frac{5!}{3!}$$

SOLUTION: Write

$$5! = 5 \cdot 4 \cdot 3 \cdot 2 \cdot 1$$

and

$$3! = 3 \cdot 2 \cdot 1$$

then

$$\frac{5!}{3!} = \frac{5 \cdot 4 \cdot 3 \cdot 2 \cdot 1}{3 \cdot 2 \cdot 1}$$

and by simplification

$$\frac{5 \cdot 4 \cdot 3 \cdot 2 \cdot 1}{3 \cdot 2 \cdot 1} = 5 \cdot 4$$

$$= 20$$

The previous example could have been solved by writing

$$\frac{5!}{3!} = \frac{3!\ 4 \cdot 5}{3!}$$

$$= 5 \cdot 4$$

Notice that we wrote

$$5! = 5 \cdot 4 \cdot 3 \cdot 2 \cdot 1$$

and combined the factors

$$3 \cdot 2 \cdot 1$$

as

$$3!$$

then

$$5! = 3!\ 4 \cdot 5$$

EXAMPLE: Find the value of

$$\frac{6! - 4!}{4!}$$

SOLUTION: Write

$$6! = 4!\ 5 \cdot 6$$

and

$$4! = 4!\,1$$

then

$$\frac{6! - 4!}{4!} = \frac{4!\,(5 \cdot 6 - 1)}{4!}$$

$$= (5 \cdot 6 - 1)$$

$$= 29$$

Notice that 4! was factored from the expression

$$6! - 4!$$

THEOREM

If n and r are positive integers, with n greater than r, then

$$n! = r!\,(r + 1)(r + 2) \cdots n$$

This theorem allows us to simplify an expression as follows:

$$5! = 4!\,5$$

$$= 3!\,4 \cdot 5$$

$$= 2!\,3 \cdot 4 \cdot 5$$

$$= 1 \cdot 2 \cdot 3 \cdot 4 \cdot 5$$

Another example is

$$(n + 2)! = (n + 1)!\,(n + 2)$$

$$= n!\,(n + 1)(n + 2)$$

$$= (n - 1)!\,n(n + 1)(n + 2)$$

EXAMPLE: Simplify

$$\frac{(n + 3)!}{n!}$$

SOLUTION: Write

$$(n + 3)! = n!\ (n + 1)(n + 2)(n + 3)$$

then

$$\frac{(n + 3)!}{n!} = \frac{n!\ (n + 1)(n + 2)(n + 3)}{n!}$$

$$= (n + 1)(n + 2)(n + 3)$$

PROBLEMS: Find the value of problems 1-4 and simplify problems 5 and 6.

1. $6!$
2. $3!\ 4!$
3. $\frac{8!}{11!}$
4. $\frac{5! - 3!}{3!}$
5. $\frac{n!}{(n - 1)!}$
6. $\frac{(n + 2)!}{n!}$

ANSWERS:

1. 720
2. 144
3. $\frac{1}{990}$
4. 19
5. n
6. $(n + 1)(n + 2)$

COMBINATIONS

As indicated previously, a combination is the selection of a certain number of objects taken from a group of objects without regard to order. We use the symbol ${}_5C_3$ to indicate that we have five objects taken three at a time, without regard to order. Using the letters A, B, C, D, and E, to designate the five objects, we list the combinations as follows:

ABC ABD ABE ACD ACE

ADE BCD BCE BDE CDE

We find there are ten combinations of five objects taken three at a time. We made the selection of three objects, as shown, but we called these selections combinations. The word combinations infers that order is not considered.

EXAMPLE: Suppose we wish to know how many color combinations can be made from four different colored marbles, if we use only three marbles at a time. The marbles are colored red, green, white, and blue.

SOLUTION: We let the first letter in each word indicate the color, then we list the possible combinations as follows:

RGW RGY RWY GWY

If we rearrange the first group, RGW, to form GWR or RWG we still have the same color combination; therefore order is not important.

The previous examples are completely within our capabilities, but suppose we have 20 boys and wish to know how many different basket ball teams we could form, one at a time, from these boys. Our listing would be quite lengthy and we would have a difficult task to determine that we had all of the possible combinations. In fact, there would be over 15,000 combinations we would have to list. This indicates the need for a formula for combinations.

FORMULA

The general formula for possible combinations of r objects from a group of n objects is

$$_{n}C_{r} = \frac{n(n-1)\ldots(n-r+1)}{1\cdot 2\cdot 3\cdots r}$$

The denominator may be written as

$$1 \cdot 2 \cdot 3 \cdots r = r!$$

and if we multiply both numerator and denominator by

$$(n - r) \cdots 2 \cdot 1$$

which is

$$(n - r)!$$

we have

$${}_nC_r = \frac{n(n - 1) \cdots (n - r + 1)(n - r) \cdots 2 \cdot 1}{r!\,(n - r) \cdots 2 \cdot 1}$$

The numerator

$$n(n - 1) \cdots (n - r + 1)(n - r) \cdots 2 \cdot 1$$

is

$$n!$$

Then

$${}_nC_r = \frac{n!}{r!\,(n - r)!}$$

This formula is read: The number of combinations of n objects taken r at a time is equal to n factorial divided by r factorial times n minus r factorial.

EXAMPLE: In the previous problem where 20 boys were available, how many different basketball teams could be formed?

SOLUTION: If the choice of which boy played center, guard, or forward is not considered, we find we desire the number of combinations of 20 boys taken five at a time and write

$${}_nC_r = \frac{n!}{r!\,(n - r)!}$$

where

$$n = 20$$

and

$$r = 5$$

Then, by substitution we have

$$_nC_r = {}_{20}C_5 = \frac{20!}{5!\,(20 - 5)!}$$

$$= \frac{20!}{5!\,15!}$$

$$= \frac{15!\ 16 \cdot 17 \cdot 18 \cdot 19 \cdot 20}{15!\ 5!}$$

$$= \frac{16 \cdot 17 \cdot 18 \cdot 19 \cdot 20}{5 \cdot 4 \cdot 3 \cdot 2 \cdot 1}$$

$$= \frac{16 \cdot 17 \cdot 3 \cdot 19 \cdot 1}{1}$$

$$= 15{,}504$$

EXAMPLE: A man has, in his pocket, a silver dollar, a half-dollar, a quarter, a dime, a nickel, and a penny. If he reaches into his pocket and pulls out three coins, how many different sums may he have?

SOLUTION: The order in not important, therefore the number of combinations of coins possible is

$$_6C_3 = \frac{6!}{3!\,(6 - 3)!}$$

$$= \frac{6!}{3!\ 3!}$$

$$= \frac{3!\ 4 \cdot 5 \cdot 6}{3!\ 3!}$$

$$= \frac{4 \cdot 5 \cdot 6}{3 \cdot 2 \cdot 1}$$

$$= \frac{4 \cdot 5}{1}$$

$$= 20$$

EXAMPLE: Find the value of

$${}_3C_3$$

SOLUTION: We use the formula given and find that

$${}_3C_3 = \frac{3!}{3!\,(3-3)!}$$

$$= \frac{3!}{3!\;0!}$$

This seems to violate the rule, "division by zero is not allowed," but we define 0! as equal 1. Then

$$\frac{3!}{3!\;0!} = \frac{3!}{3!} = 1$$

which is obvious if we list the combinations of three things taken three at a time.

PROBLEMS: Find the value of problems 1-6 and solve problems 7, 8, and 9.

1. ${}_6C_2$
2. ${}_6C_4$
3. ${}_{15}C_5$
4. ${}_7C_7$
5. $\dfrac{{}_6C_3 + {}_7C_3}{{}_{13}C_6}$
6. $\dfrac{({}_7C_3)\,({}_6C_3)}{{}_{14}C_4}$

7. We want to paint three rooms in a house, each a different color and we may choose from seven different colors of paint. How many color combinations are possible, for the three rooms?

8. If 20 boys go out for the football team, how many different teams may be formed, one at a time?

9. Two boys and their dates go to the drive-in and each wants a different flavor ice cream cone. The drive-in has 24 flavors of ice cream. How many combinations of flavors may they choose?

ANSWERS:

1. 15

2. 15

3. 3,003

4. 1

5. $\frac{5}{156}$

6. $\frac{100}{143}$

7. 35

8. 167,960

9. 10,626

PRINCIPLE OF CHOICE

The principle of choice is discussed in relation to combinations although it is also, later in this chapter, discussed in relation to permutations. It is stated as follows:

If a selection can be made in n_1 ways, and after this selection is made, a second selection can be made in n_2 ways, and after this selection is made, a third selection can be made in

n_3 ways, and so forth for r ways, then the r selections can be made together in

$$n_1 \cdot n_2 \cdot n_3 \cdots \cdot n_r \text{ ways}$$

EXAMPLE: In how many ways can a coach choose first a football team and then a basketball team if 18 boys go out for either team?

SOLUTION: First let the coach choose a football team. That is

$$\begin{aligned} {}_{18}C_{11} &= \frac{18!}{11!\,(18-11)!} \\ &= \frac{18!}{11!\,7!} \\ &= \frac{11!\; 12 \cdot 13 \cdot 14 \cdot 15 \cdot 16 \cdot 17 \cdot 18}{11!\; 7 \cdot 6 \cdot 5 \cdot 4 \cdot 3 \cdot 2 \cdot 1} \\ &= 31,824 \end{aligned}$$

The coach now must choose a basketball team from the remaining seven boys. That is

$$\begin{aligned} {}_{7}C_{5} &= \frac{7!}{5!\,(7-5)!} \\ &= \frac{7!}{5!\,2!} \\ &= \frac{5!\; 6 \cdot 7}{5!\; 2!} \\ &= \frac{6 \cdot 7}{2} \\ &= 21 \end{aligned}$$

Then, together, the two teams can be chosen in

$$(31,824)\,(21) = 668,304 \text{ ways}$$

EXAMPLE: A man ordering dinner has a choice of one meat dish from four, a choice of three vegetables from seven, one salad from three, and one dessert from four. How many different menus are possible?

SOLUTION: The individual combinations are as follows:

$$\text{meat} \ldots\ldots\ldots {}_4C_1$$

$$\text{vegetable} \ldots\ldots {}_7C_4$$

$$\text{salad} \ldots\ldots\ldots {}_3C_1$$

$$\text{dessert} \ldots\ldots {}_4C_1$$

The value of

$$
\begin{aligned}
{}_4C_1 &= \frac{4!}{1!\,(4-1)!} \\
&= \frac{4!}{3!} \\
&= 4
\end{aligned}
$$

and

$$
\begin{aligned}
{}_7C_4 &= \frac{7!}{4!\,(7-4)!} \\
&= \frac{7!}{4!\,3!} \\
&= \frac{5\cdot 6\cdot 7}{2\cdot 3} \\
&= 35
\end{aligned}
$$

and

$$
\begin{aligned}
{}_3C_1 &= \frac{3!}{1!\,(3-1)!} \\
&= \frac{3!}{2!} \\
&= 3
\end{aligned}
$$

therefore, there are

$$(4)\,(35)\,(3)\,(4) = 1680$$

different menus available to the man.

PROBLEMS: Solve the following problems.

1. A man has 12 different colored shirts and 20 different ties. How many shirt and tie

combinations can he select to take on a trip, if he takes three shirts and five ties?

2. A petty officer, in charge of posting the watch, has in the duty section 12 men. He must post three different fire watches, then post four aircraft guards on different aircraft. How many different assignments of men can he make?

3. If there are 10 third class and 14 second class petty officers in a division which must furnish two second class and six third class petty officers for shore patrol, how many different shore patrol parties can be made?

ANSWERS:

1. 3,410,880
2. 27,720
3. 19,110

PERMUTATIONS

Permutations are similar to combinations but extend the requirements of combinations by considering order.

Suppose we have two letters, A and B, and wish to know how many arrangements of these letters can be made. It is obvious that the answer is two. That is

AB and BA

If we extend this to the three letters A, B, and C, we find the answer to be

ABC, ACB, BAC, BCA, CAB, CBA

We had three choices for the first letter, and after we chose the first letter, we had only two choices for the second letter, and after the second letter, we had only one choice. This is shown in the "tree" diagram in figure 16-1. Notice that there is a total of six different paths to the ends of the "branches" of the "tree" diagram.

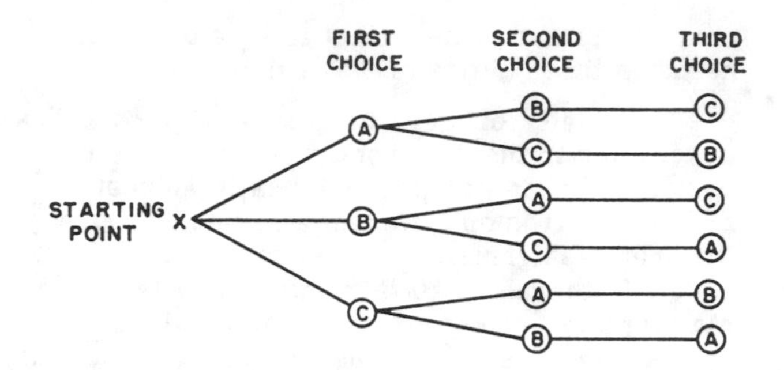

Figure 16-1.—"Tree" diagram.

If the number of objects is large, the tree would become very complicated; therefore, we approach the problem in another manner, using parentheses to show the possible choices. Suppose we were to arrange five objects in as many different orders as possible. We have for the first choice six objects.

(6) ( ) ( ) ( ) ( ) ( )

For the second choice we have only five choices.

(6) (5) ( ) ( ) ( ) ( )

For the third choice we have only four choices.

(6) (5) (4) ( ) ( ) ( )

This may be continued as follows:

(6) (5) (4) (3) (2) (1)

By applying the principle of choice we find the total possible ways of arranging the objects to be the product of the individual choices. That is

$$6 \cdot 5 \cdot 4 \cdot 3 \cdot 2 \cdot 1$$

and this may be written as

$$6!$$

This leads to the statement: The number of permutations of n objects, taken all together, is equal to n!.

EXAMPLE: How many permutations of seven different letters may be made?

SOLUTION: We could use the "tree" but this would become complicated. (Try it to see why.) We could use the parentheses as follows:

$$(7)\,(6)\,(5)\,(4)\,(3)\,(2)\,(1) = 5040$$

The easiest solution is to use the previous statement and write

$${}_7P_7 = 7!$$

That is, the number of possible arrangements of seven objects, taken seven at a time, is 7!. NOTE: Compare this with the number of COMBINATIONS of seven objects, taken seven at a time.

If we are faced with finding the number of permutations of seven objects taken three at a time, we use three parentheses as follows:

In the first position we have a choice of seven objects.

$$(7)\,(\ \ )\,(\ \ )$$

In the second position we have a choice of six objects

$$(7)\,(6)\,(\ \ )$$

In the last position we have a choice of five objects,

$$(7)\,(6)\,(5)$$

and by principle of choice, the solution is

$$7 \cdot 6 \cdot 5 = 210$$

FORMULA

At this point we will use our knowledge of combinations to develop a formula for the number of permutations of n objects taken r at a time.

Suppose we wish to find the number of permutations of five things taken three at a time. We first determine the number of groups of three, as follows:

$$ {}_5C_3 = \frac{5!}{3!\,(5-3)!} $$

$$ = \frac{5!}{3!\,2!} $$

$$ = 10 $$

Thus, there are 10 groups of three objects each.

We are now asked to arrange each of these ten groups in as many orders as possible. We know that the number of permutations of three objects, taken together, is 3!. We may arrange each of the 10 groups in 3! or six ways. The total number of possible permutations of ${}_5C_3$ then is

$$ {}_5C_3 \cdot 3! = 10 \cdot 6 $$

which is written

$$ {}_5C_3 \cdot 3! = {}_5P_3 $$

Put into the general form, then

$$ {}_nC_r \cdot r! = {}_nP_r $$

and knowing that

$$ {}_nC_r = \frac{n!}{r!\,(n-r)!} $$

then

$$ {}_nC_r \cdot r! = \frac{n!}{r!\,(n-r)!} \cdot r! $$

$$ = \frac{n!}{(n-r)!} $$

but

$$ {}_nC_r \cdot r! = {}_nP_r $$

therefore

$$ {}_nP_r = \frac{n!}{(n - r)!} $$

EXAMPLE: How many permutations of six objects, taken two at a time, can be made?

SOLUTION: The number of permutations of six objects, taken two at a time, is written

$$ {}_6P_2 = \frac{6!}{(6 - 2)!} $$

$$ = \frac{6!}{4!} $$

$$ = \frac{4!\ 5 \cdot 6}{4!} $$

$$ = 5 \cdot 6 $$

$$ = 30 $$

EXAMPLE: In how many ways can eight people be arranged in a row?

SOLUTION: All eight people must be in the row; therefore, we want the number of permutations of eight people, taken eight at a time, which is

$$ {}_8P_8 = \frac{8!}{(8 - 8)!} $$

$$ = \frac{8!}{0!} $$

(Remember that 0! was defined as equal to 1) then

$$ \frac{8!}{0!} = \frac{8 \cdot 7 \cdot 6 \cdot 5 \cdot 4 \cdot 3 \cdot 2 \cdot 1}{1} $$

$$ = 40,320 $$

APPENDIX

I TRIGONOMETRIC FORMULAS

$$\sin\theta \csc\theta = \cos\theta \sec\theta = \tan\theta \cot\theta = 1$$

$$\tan\theta = \frac{\sin\theta}{\cos\theta}, \qquad \cot\theta = \frac{\cos\theta}{\sin\theta}$$

$$\sec^2 = 1 + \tan^2\theta, \qquad \csc^2\theta = 1 + \cot^2\theta$$

$$\sin^2\theta + \cos^2\theta = 1$$

$$\sin(\theta \pm \phi) = \sin\theta \cos\phi \pm \cos\theta \sin\phi$$

$$\cos(\theta \pm \phi) = \cos\theta \cos\phi \mp \sin\theta \sin\phi$$

$$\tan(\theta \pm \phi) = \frac{\tan\theta \pm \tan\phi}{1 \mp \tan\theta \tan\phi}$$

$$\cot(\theta \pm \phi) = \frac{\cot\theta \cot\phi \mp 1}{\cot\phi \pm \cot\theta}$$

$$\sin 2\theta = 2\sin\theta \cos\theta$$

$$\sin 3\theta = 3\sin\theta - 4\sin^3\theta$$

$$\sin 4\theta = 8\cos^3\theta \sin\theta - 4\cos\theta \sin\theta$$

$$\cos 2\theta = 2\cos^2\theta - 1 = \cos^2\theta - \sin^2\theta$$

$$\cos 3\theta = 4\cos^3\theta - 3\cos\theta$$

$$\cos 4\theta = 8\cos^4\theta - 8\cos^2\theta + 1$$

$$\sin\theta \pm \sin\phi = 2\sin\tfrac{1}{2}(\theta \pm \phi)\cos\tfrac{1}{2}(\theta \pm \phi)$$

$$\cos\theta + \cos\phi = 2\cos\tfrac{1}{2}(\theta + \phi)\cos\tfrac{1}{2}(\theta - \phi)$$

$$\cos\theta - \cos\phi = -2\sin\tfrac{1}{2}(\theta + \phi)\sin\tfrac{1}{2}(\theta - \phi)$$

$$\tan\theta \pm \tan\phi = \frac{\sin(\theta \pm \phi)}{\cos\theta \cos\phi} , \quad \cot\theta \pm \cot\phi = \pm \frac{\sin(\theta \pm \phi)}{\sin\theta \sin\phi}$$

$$\sin\frac{\theta}{2} = \pm\left(\frac{1-\cos\theta}{2}\right)^{\frac{1}{2}}, \quad \cos\frac{\theta}{2} = \pm\left(\frac{1+\cos\theta}{2}\right)^{\frac{1}{2}}$$

$$\tan\frac{\theta}{2} = \frac{1-\cos\theta}{\sin\theta} = \frac{\sin\theta}{1+\cos\theta} = \pm\left(\frac{1-\cos\theta}{1+\cos\theta}\right)^{\frac{1}{2}}$$

$$\frac{\sin\theta \pm \sin\phi}{\cos\theta + \cos\phi} = \tan\tfrac{1}{2}(\theta \pm \phi)$$

Sine law:

$$\frac{a}{\sin\theta} = \frac{b}{\sin\phi} = \frac{c}{\sin\Psi}$$

Cosine law:

$$a^2 = b^2 + c^2 - 2bc\cos\theta$$

$$b^2 = c^2 + a^2 - 2ca\cos\phi$$

$$c^2 = a^2 + b^2 - 2ab\cos\Psi$$

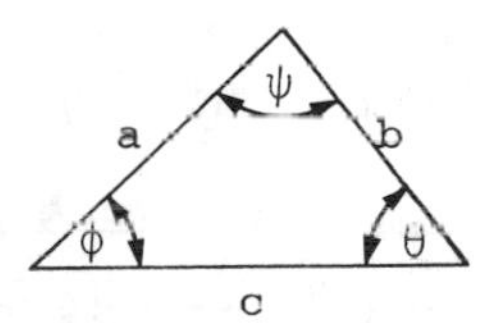

II INDEFINITE INTEGRALS

1. $\int x^n dx = \frac{x^{n+1}}{n+1}, \; n \neq -1$

2. $\int \frac{dx}{x} = \ln |x|$

3. $\int a^x dx = \frac{a^x}{\ln a}$

4. $\int \tan x \, dx = -\ln |\cos x|$

5. $\int \cot x \, dx = \ln |\sin x|$

6. $\int \sec x \, dx = \ln |\sec x + \tan x|$

7. $\int \frac{dx}{x-a} = \ln |x-a|$

8. $\int \frac{dx}{x^2+a^2} = \frac{1}{a} \tan^{-1} \frac{x}{a}$

9. $\int \frac{x\,dx}{x^2+a^2} = \frac{1}{2} \ln (x^2+a^2)$

10. $\int \frac{dx}{x^2-a^2} = \frac{1}{2a} \ln \left| \frac{x-a}{x+a} \right|$

11. $\int \frac{dx}{(a^2-x^2)^{\frac{1}{2}}} = \sin^{-1} \frac{x}{a}$

12. $\int \frac{dx}{(x^2 \pm a^2)^{\frac{1}{2}}} = \ln |x+(x^2 \pm a^2)^{\frac{1}{2}}|$

13. $\int (a^2-x^2)^{\frac{1}{2}} dx = \frac{1}{2} \left[x(a^2-x^2)^{\frac{1}{2}} + a^2 \sin^{-1} \frac{x}{a} \right]$

14. $\int x(a^2-x^2)^{\frac{1}{2}}dx = -1/3(a^2-x^2)^{\frac{3}{2}}$

15. $\int (x^2 \pm a^2)^{\frac{1}{2}}dx = \frac{1}{2}x(x^2 \pm a^2)^{\frac{1}{2}} \pm \frac{a^2}{2} \ln |x+(x^2 \pm a^2)^{\frac{1}{2}}|$, $a > 0$

16. $\int \frac{dx}{(a^2-x^2)^{3/2}} = \frac{x}{a^2(a^2-x^2)^{\frac{1}{2}}}$

17. $\int \frac{xdx}{(a^2 \pm x^2)^{\frac{1}{2}}} = \pm(a^2 \pm x^2)^{\frac{1}{2}}$

18. $\int x(x^2 \pm a^2)^{\frac{1}{2}}dx = 1/3(x^2 \pm a^2)^{3/2}$

19. $\int \frac{dx}{(x^2 \pm a^2)^{3/2}} = \frac{\pm x}{a^2(x^2 \pm a^2)^{\frac{1}{2}}}$

20. $\int \sin ax\, dx = -\frac{1}{a} \cos ax$

21. $\int \sin^2 ax\, dx = \frac{1}{2a}(ax - \frac{1}{2}\sin 2ax)$

22. $\int \cos ax\, dx = \frac{1}{a} \sin ax$

23. $\int \cos^2 ax dx = \frac{1}{2a}(ax + \frac{1}{2}\sin 2ax)$

24. $\int \sec^2 x\, dx = \tan x$

25. $\int \sec x \tan x\, dx = \sec x$

26. $\int \sin ax \sin bx dx = \frac{\sin(a-b)x}{2(a-b)} - \frac{\sin(a+b)x}{2(a+b)}$, $a \neq b$

27. $\int \cos ax \cos bx dx = \frac{\sin(a-b)x}{2(a-b)} + \frac{\sin(a+b)x}{2(a+b)}$, $a \neq b$

28. $\int e^{ax}dx = \frac{e^{ax}}{a}$

29. $\int xe^{ax}dx = \frac{xe^{ax}}{a^2} - \frac{e^{ax}}{a^2}$

30. $\int \ln x \, dx = x(\ln x - 1)$

31. $\int x \ln x \, dx = x^2\left(\frac{\ln x}{2} - \frac{1}{4}\right)$

32. $\int \sinh ax dx = \frac{1}{a} \cosh ax$

33. $\int \cosh ax dx = \frac{1}{a} \sinh ax$

34. $\int \tan^3 x \, dx = \frac{1}{2}\tan^2 x + \ln \cos x$

35. $\int \cot^3 x \, dx = -\frac{1}{2}\cot^2 x - \ln \sin x$

36. $\int \tan^2 x \, dx = \tan x - x$

37. $\int \cot^2 x \, dx = -\cot x - x$

38. $\int x \sin x \, dx = \sin x - x \cos x$

39. $\int x^2 \sin x \, dx = 2x \sin x - (x^2 - 2)\cos x$

40. $\int x^3 \sin x \, dx = (3x^2 - 6)\sin x - (x^3 - 6x)\cos x$

41. $\int x \cos x \, dx = \cos x + x \sin x$

42. $\int x^2\cos x\, dx = 2x \cos x + (x^2-2)\sin x$

43. $\int x^3\cos x\, dx = (3x^2-6)\cos x + (x^3-6x)\sin x$

44. $\int \sin^{-1}x\, dx = x \sin^{-1}x + (1-x^2)^{\frac{1}{2}}$

45. $\int \cos^{-1}x\, dx = x \cos^{-1}x - (1-x^2)^{\frac{1}{2}}$

46. $\int \tan^{-1}x\, dx = x \tan^{-1}x - \frac{1}{2} \ln(1+x^2)$

47. $\int \cot^{-1}x\, dx = x \cot^{-1}x + \frac{1}{2} \ln(1+x^2)$

48. $\int \sec^{-1}x\, dx = x \sec^{-1}x - \ln [x+(x^2-1)^{\frac{1}{2}}]$

49. $\int \csc^{-1}x\, dx = x \csc^{-1}x + \ln [x+(x^2-1)^{\frac{1}{2}}]$

50. $\int x \sin^{-1}x\, dx = \frac{1}{4}[(2x^2-1)\sin^{-1}x+x(1-x^2)^{\frac{1}{2}}]$

51. $\int x \cos^{-1}x\, dx = \frac{1}{4}[(2x^2-1)\cos^{-1}x-x(1-x^2)^{\frac{1}{2}}]$

52. $\int x \tan^{-1}x\, dx = \frac{1}{2}(x^2+1)\tan^{-1}x - \frac{x}{2}$

53. $\int (\sin^{-1}x)^2\, dx = x(\sin^{-1}x)^2 - 2x+2(1-x^2)^{\frac{1}{2}}\sin^{-1}x$

54. $\int (\cos^{-1}x)^2 dx = x(\cos^{-1}x)^2 - 2x-2(1-x^2)^{\frac{1}{2}} \cos^{-1}x$

55. $\int \sin(\ln x)dx = \frac{1}{2}x \sin(\ln x) - \frac{1}{2}x \cos(\ln x)$

56. $\int \cos(\ln x)dx = \frac{1}{2}x \sin(\ln x) + \frac{1}{2}x \cos(\ln x)$

57. $\int \tanh x \, dx = \ln \cosh x$

58. $\int \coth x \, dx = \ln \sinh x$

59. $\int \operatorname{sech} x \, dx = \tan^{-1}(\sinh x)$

60. $\int \operatorname{csch} x \, dx = \ln \left(\tanh \frac{x}{2}\right)$

61. $\int \frac{dx}{x(a+bx)^{\frac{1}{2}}} = \frac{1}{a^{\frac{1}{2}}} \ln \frac{(a+bx)^{\frac{1}{2}}-a^{\frac{1}{2}}}{(a+bx)^{\frac{1}{2}}+a^{\frac{1}{2}}}$

62. $\int \frac{xdx}{a+bx} = \frac{x}{b} - \frac{a}{b^2} \ln(a+bx)$

63. $\int \frac{xdx}{(a+bx)^2} = \frac{1}{b^2}\left[\ln(a+bx) + \frac{a}{a+bx}\right]$

64. $\int \frac{x^2 dx}{a+bx} = \frac{1}{b^3}[\frac{1}{2}(a+bx)^2 - 2a(a+bx) + a^2 \ln(a+bx)]$

65. $\int \frac{dx}{x(a+bx)} = -\frac{1}{a} \ln \frac{a+bx}{x}$

66. $\int \frac{dx}{x(a+bx)^2} = \frac{1}{a(a+bx)} - \frac{1}{a^2} \ln \frac{a+bx}{x}$

67. $\int \frac{dx}{a+be^{rx}} = \frac{x}{a} - \frac{1}{ar} \ln(a+be^{rx})$

68. $\int e^{ax} \ln x \, dx = \frac{e^{ax} \ln x}{a} - \frac{1}{a} \int \frac{e^{ax}}{x} dx$

69. $\int e^{ax} \sin bx \, dx = e^{ax} \frac{a \sin bx - b \cos bx}{a^2 + b^2}$

70. $\int e^{ax} \cos bx \, dx = e^{ax} \frac{a \cos bx + b \sin bx}{a^2 + b^2}$

71. $\int \frac{dx}{a+bx^2} = \begin{cases} \frac{1}{(ab)^{\frac{1}{2}}} \tan^{-1} \frac{x(ab)^{\frac{1}{2}}}{a} & \text{if } ab > 0; \\ \frac{1}{(-ab)^{\frac{1}{2}}} \tanh^{-1} \frac{x(-ab)^{\frac{1}{2}}}{a} & \text{if } ab < 0 \end{cases}$

72. $\int \frac{dx}{a^2+b^2x^2} = \frac{1}{ab} \tan^{-1} \frac{bx}{a}$

73. $\int \frac{xdx}{a+bx^2} = \frac{1}{2b} \ln(a+bx^2)$

III DEFINITE INTEGRALS

1. $$\int_0^{\infty} x^n e^{-ax} dx = \frac{n!}{a^{n+1}}$$

2. $$\int_1^{\infty} \frac{dx}{x^a} = \frac{1}{a-1}, \quad a > 1$$

3. $$\int_0^{\infty} \frac{dx}{(1+x)x^a} = \pi \csc a\pi, \quad 0 < a < 1$$

4. $$\int_0^{\infty} \frac{x^{a-1}}{1+x} dx = \frac{\pi}{\sin a\pi}, \quad 0 < a < 1$$

5. $$\int_0^{\infty} \frac{a\,dx}{a^2+x^2} = \begin{cases} \frac{\pi}{2} & \text{if } a > 0 \\ 0 & \text{if } a = 0 \\ -\frac{\pi}{2} & \text{if } a < 0 \end{cases}$$

6. $$\int_0^{\pi/2} \cos^n x\, dx = \int_0^{\pi/2} \sin^n x\, dx = \begin{cases} \frac{1 \cdot 3 \cdot 5 \cdots (n-1)}{2 \cdot 4 \cdot 6 \cdots n} \frac{\pi}{2} & \text{for } n \text{ even} \\ \frac{2 \cdot 4 \cdot 6 \cdots (n-1)}{1 \cdot 3 \cdot 5 \cdots n} & \text{for } n \text{ odd} \end{cases}$$

7. $$\int_0^{\infty} \frac{\sin x}{x} dx = \frac{\pi}{2}$$

8. $$\int_0^{\infty} \frac{\cos ax}{1+x^2} dx = \frac{\pi}{2} e^{-|a|}$$

9. $\int_0^{\pi} \sin nx \sin mx \, dx = \int_0^{\pi} \cos nx \cos mx dx = 0$

m,n are integers and m≠n

10. $\int_0^{\pi} \sin^2 nx dx = \int_0^{\pi} \cos^2 nx \, dx = \frac{\pi}{2}$

11. $\int_0^{\infty} \frac{\sin^2 x}{x} \, dx = \frac{\pi}{2}$

12. $\int_0^{\infty} \sin^2 x \, dx = \int_0^{\infty} \cos^2 x \, dx = \frac{1}{2}\left(\frac{\pi}{2}\right)^{\frac{1}{2}}$

13. $\int_0^{\infty} \frac{\sin x}{x^{\frac{1}{2}}} \, dx = \int_0^{\infty} \frac{\cos x}{x^{\frac{1}{2}}} \, dx = \left(\frac{\pi}{2}\right)^{\frac{1}{2}}$

14. $\int_0^{\infty} e^{-ax} \, dx = \frac{1}{a}$, $a > 0$

15. $\int_0^{\infty} \frac{e^{-ax} - e^{-bx}}{x} \, dx = \ln \frac{b}{a}$ $a, b > 0$

16. $\int_0^{\infty} e^{-a^2x^2} dx = \frac{1}{2a} \pi^{\frac{1}{2}}$

17. $\int_0^{\infty} xe^{-x^2} \, dx = \frac{1}{2}$

18. $\int_0^{\infty} x^2 e^{-x^2} dx = \frac{\pi^{\frac{1}{2}}}{4}$

19. $\int_0^{\infty} x^2 e^{-ax^2} dx = \frac{\pi^{\frac{1}{2}}}{4} a^{-3/2}$

20. $\int_0^{\infty} e^{-a^2x^2} \cos x \, dx = \frac{\pi}{2|a|} \quad e^{-1/4a^2}, \; a \neq 0$

21. $\int_0^{\infty} x[\mathrm{erf}(ax)] e^{-b^2x^2} dx = \frac{a}{2b^2} (a^2+b^2)^{-\frac{1}{2}}$

22. $\int_0^{\infty} x^2[\mathrm{erf}(ax)] e^{-b^2x^2} dx = \frac{1}{2\pi^{\frac{1}{2}}} \left[\frac{a}{b^2(a^2+b^2)} + \frac{1}{b^3} \tan^{-1} \frac{a^2}{b} \right]$

23. $\int_0^{\infty} [\mathrm{erf}(ax)] e^{-b^2x^2} dx = \frac{1}{(\pi b^2)^{\frac{1}{2}}} \tan^{-1} \frac{a}{b}$

24. $\int_0^{\infty} [1-\mathrm{erf}(ax)] dx = \frac{1}{a\,\pi^{\frac{1}{2}}}$

25. $\int_0^{1} (\ln x)^n dx = (-1)^n n!$

26. $\int_0^{1} \frac{\ln x}{1+x} dx = \frac{\pi^2}{12}$

27. $\int_0^{1} \frac{\ln x}{(1-x^2)^{\frac{1}{2}}} dx = -\frac{\pi}{2} \ln 2$

28. $\int_0^{\pi/2} \ln(\tan x) dx = 0$

29. $\int_0^{\pi} x \ln(\sin x) dx = -\frac{\pi^2}{2} \ln 2$